MW00761080

Springer
Berlin
Heidelberg
New York
Hong Kong
London
Milan
Paris
Tokyo

Physics and Astronomy

ONLINE LIBRARY

http://www.springer.de/phys/

ADVANCES IN MATERIALS RESEARCH

Series Editor-in-Chief: Y. Kawazoe

Series Editors: M. Hasegawa   A. Inoue   N. Kobayashi   T. Sakurai   L. Wille

The series Advances in Materials Research reports in a systematic and comprehensive way on the latest progress in basic materials sciences. It contains both theoretically and experimentally oriented texts written by leading experts in the field. Advances in Materials Research is a continuation of the series Research Institute of Tohoku University (RITU).

Series homepage – http://www.springer.de/phys/books/amr/

Y. Kawazoe   Y. Waseda (Eds.)

Structure and Properties of Aperiodic Materials

With 145 Figures

Springer

Professor Yoshiyuki Kawazoe
Institute for Materials Research
Tohoku University
2-1-1 Katahira, Aoba-ku
Sendai 980-8577, Japan
e–mail: kawazoe@imr.edu

Professor Yoshio Waseda
Institute of Multidisciplinary Reserach
for Advanced Materials
Tohoku University
2-1-1 Katahira, Aoba-ku
Sendai 980-8577, Japan
e–mail: waseda@tagen.tohoku.ac.jp

Series Editor-in-Chief:

Professor Yoshiyuki Kawazoe
Institute for Materials Research, Tohoku University
2-1-1 Katahira, Aoba-ku, Sendai 980-8577, Japan

Series Editors:

Professor Masayuki Hasegawa
Professor Akihisa Inoue
Professor Norio Kobayashi
Professor Toshio Sakurai
Institute for Materials Research, Tohoku University
2-1-1 Katahira, Aoba-ku, Sendai 980-8577, Japan

Professor Luc Wille
Department of Physics, Florida Atlantic University
777 Glades Road, Boca Raton, FL 33431, USA

Library of Congress Cataloging-in-Publication Data: Structure and properties of aperiodic materials/ Y. Kawazoe and Y. Waseda (eds.). p. cm. – (Advances in materials research, ISSN 1435-1889; 5) Includes bibliographical references and index. ISBN 3-540-00959-0 (alk. paper) 1. Crystallography, Mathematical. 2 Aperiodicity. 3. Amorphous substances. I. Kawazoe, Y. (Yoshiyuki), 1947–. II. Waseda, Yoshio. III. Series. QD911.S77 2003 548'.7–dc21 2003050422

ISSN 1435-1889
ISBN 3-540-00959-0 Springer-Verlag Berlin Heidelberg New York

Typesetting: Data conversion by LE-T$_E$X, Leipzig
Cover concept: eStudio Calamar Steinen
Cover design: *design & production*, Heidelberg

Printed on acid-free paper   SPIN 10893023   57/3141/ba - 5 4 3 2 1 0

Series Preface by the Editor-in-Chief

The present book is the fifth volume of the Springer Series 'Advances in Materials Research', which is edited mainly by staff scientists from the Institute for Materials Research (IMR or Kin-Ken in Japanese) of Tohoku University. The series began with the book 'Microscopic Dynamics of Fracture', and has already been followed by three books.

Further volumes are planned from staff scientists joining the Institute of Multidisciplinary Research for Advanced Materials (IMRAM or Tagen-Ken in Japanese), which commenced its activities in April 2001. This new integral organization for materials research in Tohoku University is expected to foster various subjects on advanced materials and processes through synergetic interactions between staff members from different disciplines. Together these books will provide high-quality reviews of topical subjects from today's very active and broad field of materials science. The scope and aim of this fifth volume is 'Structure and Properties of Aperiodic Materials'.

As the series editor, I would like to express sincere thanks to Dr. Claus Ascheron of Springer-Verlag, who visits us often, for his kind and enlightening encouragement and continuing support of this series.

Sendai, March 2003   *Yoshiyuki Kawazoe*

Preface

In this volume we aim to introduce recent progress in the study of aperiodic materials, which include icosahedral clusters, amorphous metals, quasicrystals, glasses, and liquids. Quasicrystals, discovered in 1984, correspond to a kind of revolution in our understanding of crystallography, wherein the five-fold rotational symmetry was prohibited in long-range ordered systems.

Various interesting physicochemical properties of these materials strongly depend on structural inhomogeneity at the microscopic level, and the small angle X-ray scattering method is widely used to analyze such structures. These new materials provide fundamental improvements to materials properties, and are not only scientifically interesting but also industrially important for applications such as ultrafine magnetic recording media and future electronic devices.

This book contains three chapters. The first chapter, written by H. Tanaka and T. Fujiwara, deals with 'Electronic Structure in Aperiodic Materials', and reviews the application of theoretical methods to determine the electronic structures and resulting properties of amorphous metals, quasicrystals, and liquids. The second chapter, written by Y. Waseda, K. Sugiyama, and A.H. Shinohara, covers the recent topic of 'Anomalous Small Angle X-ray Scattering for Structural Inhomogeneity of Materials', starting with its fundamentals. The third chapter, 'Icosahedral Clusters in $RE(TM_{1-x}Al_x)_{13}$ Amorphous Alloys', by K. Fukamichi, A. Fujita, T.H. Chang, E. Matsubara, and Y. Waseda, contains numerous important results on the relationship between structural disorder and electron correlation, with special reference to their magnetic properties. This includes a unified model of the itinerant- and localized-electron models by including the concept of spin fluctuation, which successfully explains various magnetic properties in crystalline homogeneous systems.

This book provides a discussion of current progress and an up-to-date evaluation of the structure and electronic and magnetic properties of aperiodic materials. With a survey of related references, illustrations and tables, this book acts as a guide for specialists and even non-specialists who wish to become acquainted with materials research.

We would like to thank all the authors and Springer-Verlag, who completed publication of this book in an extraordinarily quick manner.

Sendai,
March 2003

Y. Kawazoe
Y. Waseda

Contents

List of Contributors

Te-Hsuan Chiang
Department
of Industrial Management
Chung Hua University
30 Tung Shiang, Hsin, Chu,
Taiwan 30012, R.O. China
tehsuan@ms18.hinet.net

Kazuaki Fukamichi
Department of Materials Science
Graduate School of Engineering
Tohoku University
Sendai 980-8579, Japan
fukamich@material.tohoku.ac.jp

Asaya Fujita
Department of Materials Science
Graduate School of Engineering
Tohoku University
Sendai 980-8579, Japan
afujita@material.tohoku.ac.jp

Takeo Fujiwara
Department
of Applied Physics
Graduate School of Engineering
The University of Tokyo
Tokyo 113-8656, Japan
fujiwara@coral.t.u-tokyo.ac.jp

Yoshiyuki Kawazoe
Institute for Materials Research
Tohoku University
Sendai 980-8577, Japan
kawazoe@imr.edu

Eiichiro Matsubara
Institute for Materials Research
Tohoku University
Sendai 880-8577, Japan
matubara@imr.tohoku.ac.jp

Armando H. Shinohara
Department
of Mechanical Engineering
Federal University of Pernambuco
Av. Academico Helio Ramos s/n
50740-530, CDU
Recife-PE, Brazil
shinohara@demec.ufpe.br

Kazumasa Sugiyama
Department of Earth
and Planetary Science
Graduate School of Science
The University of Tokyo
Tokyo 113-0033, Japan
kazumasa@eps.s.u-tokyo.ac.jp

Hiroshi Tanaka
Department of Materials Science
Faculty of Science and Engineering
Shimane University
Matsue 690-8504, Japan
tanakah@riko.shimane-u.ac.jp

Yoshio Waseda
Institute
of Multidisciplinary Research
for Advanced Materials
Tohoku University
Sendai 980-8577, Japan
waseda@tagen.tohoku.ac.jp

1 Electronic Structure in Aperiodic Systems

H. Tanaka and T. Fujiwara

1.1 Introduction

Liquids are equilibrium systems which represent a longstanding problem for physics and chemistry. We now know a great deal about their structures and properties in conditions without large fluctuations, and the physics of liquids near the critical point has now been elucidated. Proteins and DNA in water are the current target of physics. The physics and chemistry of amorphous metals have made much progress during the last thirty years and they are now widely used in industry. Quasicrystals were discovered in 1984 and this brought about a kind of revolution in our understanding of materials: until then, every textbook had stated in the first few pages that the five-fold rotational symmetry was prohibited in long-range ordered systems. We now know that this is not the case. Furthermore, quite a few quasicrystalline systems are in an equilibrium phase.

During the last few decades, many new experimental techniques have been invented and developed, e.g., synchrotron X-ray radiation and neutron scattering experiments. As a consequence, the static structures and dynamics in these aperiodic systems are fairly well understood.

This progress has not only affected the experimental tools, but also the theories. First-principles electronic structure theories have been considerably developed over the last thirty-five years, after development of the density functional theory [1]. With the rapid development of computer technology, the theory of computation has also made much progress, e.g., the energy-linearized wavefunction method [2], the first-principles pseudopotential method [3,4], and the first-principle molecular dynamics simulation technique (Car–Parrinello method) [5]. These new theories have also prompted progress in our understanding of aperiodic systems.

In this chapter, we give a review of electronic structures in amorphous metals, quasicrystals, and liquids. In Sect. 1.2, electronic structures in amorphous metals are discussed. The Bloch theorem is not applicable in the electronic structure calculation and special calculational techniques in real space must be developed. Several amorphous metals and metallic compounds will be described. Magnetism is one of the most interesting properties in amorphous metals and several specific properties such as those of spin glasses are described. In Sect. 1.3, the electronic structures and transport properties in quasicrystals are discussed. Although the Bloch theorem cannot be used in quasicrystals, crystalline compounds exist with very similar composition and local structure to quasicrystals. A great deal of progress has thereby been obtained in the study of these compounds, called crystalline approximants. Transport is anomalous in quasicrystals. The weak localization theory will not be explained in detail here, but a basic review can be found in the literature [6]. Section 1.4 is devoted to electronic structures in liquids. The Car–Parrinello method is applied to liquid systems. This method demonstrates the importance of computational algorithms.

1.2 Amorphous Metals

1.2.1 Model Atomic Structure

In amorphous and liquid systems, it is difficult to determine the position of each atom experimentally, due to lack of translational symmetry. X-ray experimental techniques [7, 8] only give us the pair distribution functions (PDF) $g_{\alpha\beta}(r)$, which are defined as follows:

$$g_{\alpha\beta}(r) = \rho_{\alpha\beta}(r)/\rho_\beta , \qquad (1.1)$$

where $\rho_{\alpha\beta}(r)$ is the density of β atoms at a distance r from an α atom, and ρ_β is the density of the β atom. To evaluate electronic structures of amorphous and liquid systems, it is therefore important to construct a reliable atomic structure model which is consistent with experimental data.

Car and Parrinello developed an efficient algorithm for ab initio molecular dynamics (MD) calculations [5]. In ab initio MD, interatomic forces are calculated directly from electronic structure without any empirical parameters, and both the electronic and atomic structures are obtained self-consistently. Self-consistency is particularly important for systems with high covalency, such as amorphous semiconductors. Recent progress in ab initio MD will be described in Sect. 1.4. The method is suitable for disordered systems consisting of simple metals and/or semiconductors, because it is most efficient when combined with the pseudo-potential technique and plane-wave orbitals at this stage. Some empirical methods must be used when constructing an atomic structure model of an amorphous alloy including atoms such as transition metals and rare-earth atoms.

Fortunately, the covalency of these materials is weak enough to assume that atoms are configured randomly and uniformly. We can then construct an atomic structure model of the amorphous metal using relatively simple methods. The simplest model for amorphous metals and alloys is the dense random packing of hard spheres (DRPHS) model. The model was first proposed by Bernal [9,10]. He constructed the model by squeezing and kneading rubber bladders filled with ball bearings of the same size. He found that the

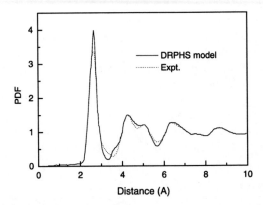

Fig. 1.1. Pair-distribution function for computer-generated amorphous iron

greater part of the DRPHS model consists of only the five types of polyhedra (so-called 'Bernal holes'). Since then, several variations of the DRPHS model have been proposed and constructed via computer simulations [7, 11, 12]. Among them, the relaxed DRPHS model developed by Cargil provides a reliable model atomic structure of amorphous alloys [7]. In this method, an atomic configuration is relaxed by interatomic forces obtained from empirical pair potentials.

Figure 1.1 shows an example PDF for amorphous Fe, calculated by Yamamoto and Doyama [13] using the relaxed DRPHS model. The first peak is sharp and clearly separated from the second and third peaks. This means that there exists a well-defined nearest-neighbor (NN) shell even in amorphous systems. The second peak splits into two peaks and their relative amplitudes also reproduce the experimental data well.

Recent progress in computing enables us to simulate the formation of amorphous or liquid alloys consisting of hundreds of atoms using empirical molecular dynamics (MD). In the relaxed DRPHS or empirical MD, it is important to construct appropriate interatomic potentials. For highly covalent systems, several multibody potentials have been proposed to describe their anisotropic bond order. On the other hand, the bond order of metallic amorphous systems seems spherical enough to describe it using pair potentials. Hafner et al. proposed a semi-empirical pair potential for transition metals by combining the pseudopotential method with a tight-binding bond-order approach [14]. This gives us a prescription for constructing a pair potential between different kinds of atoms in a semi-empirical way.

1.2.2 Calculation Scheme for Large Disordered Systems

In order to evaluate the ab initio electronic structures of large disordered systems without translational symmetry, the Hamiltonian must be described in real space. Furthermore, it is desirable to describe it using a localized basis. The tight-binding linear muffin-tin orbital (TB LMTO) is a promising method for this purpose. It provides an ab initio tight-binding Hamiltonian within the local density functional approximation (LDA) [15,16]. In this section, the TB LMTO method is explained briefly.

To calculate large systems, an efficient scheme is also necessary, evaluating the electronic structure without direct diagonalization of the Hamiltonian, because computer resources are limited. For this purpose, several schemes have been proposed. They are also introduced in this section.

TB Linear Muffin-Tin Orbit

In this section we explain the linear muffin-tin orbital (LMTO) method [2,17], and how it is transferred into a tight-binding form. In the LMTO method, the whole space is divided into muffin-tin spheres near the nuclei and in the interstitial region outside the spheres. In a muffin-tin sphere, we can approximate the potential with a spherically symmetric one, and reduce the Schrödinger equation to a radial differential equation. The basis functions are obtained by solving the Schrödinger equation within the muffin-tin sphere.

In the interstitial region, the potential is shallow and slowly varying. Therefore, we can approximate the basis functions with a linear combination of solutions for the Laplace equation. The electronic structure is determined by the condition that the wave functions within muffin-tin spheres are smoothly (i.e., continuously and differentiably) connected with those in the interstitial region on the sphere surface.

Furthermore, the basis function is expanded in the Taylor series around a certain energy $E_{\nu l}$, and its energy dependence is taken into account up to the first order, so that the secular equation is given as a linear eigenvalue problem,

$$\boldsymbol{H}\varphi_n = E_n\varphi_n .$$ (1.2)

Hereafter, we employ the atomic sphere approximation (ASA), in which the atomic sphere centered at $\boldsymbol{R}$ has the same volume as that of the Wigner–Seitz cell. As a result, atomic spheres overlap each other to some extent. However, this approximation is valid provided that the overlap of the MT spheres does not exceed 20% [18]. Then the basis function is given as

$$\chi_{\boldsymbol{R}L}(\boldsymbol{r} - \boldsymbol{R}) = \phi_{\boldsymbol{R}L}(\boldsymbol{r} - \boldsymbol{R}) + \sum_{\boldsymbol{R}'}\sum_{L'}\dot{\phi}_{\boldsymbol{R}'L'}(\boldsymbol{r} - \boldsymbol{R}')h_{\boldsymbol{R}'L',\boldsymbol{R}L} ,$$ (1.3)

where L denotes a set of quantum numbers $\{l, m\}$. The function $\phi_{\boldsymbol{R}L}(\boldsymbol{r} - \boldsymbol{R})$ is the solution of the Schrödinger equation at a certain energy $E_{\nu l}$,

$$(H - E_{\nu l})\phi_{\boldsymbol{R}L}(\boldsymbol{r} - \boldsymbol{R}) = 0 ,$$ (1.4)

within the muffin-tin centered at R. The function $\dot{\phi}_{RL}(r - R)$ is the energy derivative of $\phi_{RL}(r - R)$ at $E = E_{\nu l}$. In a physical sense, the second term on the right-hand side of (1.3) describes the tails of muffin-tin orbitals at neighboring sites centered at R'. For convenience, we rewrite (1.3) as

$$|\chi\rangle^\infty = |\phi\rangle + |\dot{\phi}\rangle h , \tag{1.5}$$

In order to construct a set of localized basis functions, we introduce a more generalized basis set defined as

$$|\bar{\chi}\rangle^\infty = |\phi\rangle + |\dot{\bar{\phi}}\rangle \bar{h} \quad \text{with} \quad |\dot{\bar{\phi}}\rangle = |\dot{\phi}\rangle + |\phi\rangle \bar{o} , \tag{1.6}$$

so that the amplitudes of tails ($\bar{h}$) decay much faster than h in (1.5) as a function of $|R - R'|$.

The coefficient h in (1.5) or $\bar{h}$ and $\bar{o}$ in (1.6) are determined according to the requirement that the basis function in (1.5) or (1.6) should be continuously and differentiably connected to the envelope function described below. The basis function in (1.5) is a special case of that in (1.6) with $\bar{o} = 0$.

In the LMTO method, the envelope function is constructed by a linear combination of solutions of the Laplace equation $\Delta\phi(r) = 0$ as

$$K_{RL}^\infty = K_{RL} - \sum_{R'L'} J_{R'L'} S_{R'L',RL} \quad \text{or} \quad |K\rangle^\infty = |K\rangle - |J\rangle S , \tag{1.7}$$

where $K_{RL}^\infty(r - R) = (|r - R|/s_w)^{(-l-1)} Y_{lm}(\widehat{r - R})$ extends over all space $\{r\}$, whereas $K_{RL}(r - R)$ and $J_{RL}(r - R)$ vanish outside their own atomic sphere. The function $K_{RL}(r - R)$ is an irregular solution of the Laplace equation with the pole at R, which decays proportionally to $|r - R|^{-l-1}$. On the other hand, $J_{RL}(r - R)$ is a regular solution of the Laplace equation with the pole at infinity, and is proportional to $|r - R|^l$. The matrix $S_{R'L',RL}$ is the so-called canonical structure constant, and is determined solely by the atomic structure. In order to obtain a localized basis, we introduce a more generalized envelope function by adding a certain amount of the irregular solution $|K\rangle$ to the regular solution $|J\rangle$ in (1.7),

$$|\bar{J}\rangle = |J\rangle - |K\rangle \bar{Q} , \tag{1.8}$$

where $\bar{Q}$ is an arbitrary diagonal matrix.

By surrounding an irregular solution K_{RL} centered at R with irregular solutions $K_{R'L'}$ centered at neighboring atoms, we can obtain a short-range envelope function, and thus a localized basis function. The new envelope function is described in the form

$$|\bar{K}\rangle^\infty = |K\rangle^\infty (I - \bar{Q}S)^{-1} = |K\rangle - |\bar{J}\rangle \bar{S} , \tag{1.9}$$

where $\bar{S}$ is given by

$$\bar{S} = S(I - \bar{Q}S)^{-1} , \quad \text{or} \quad \bar{S} = S + \bar{Q}\bar{S} . \tag{1.10}$$

We can choose an arbitrary value for either $\bar{o}$ or $\bar{Q}$ in the above equations. If we choose $\bar{o} = 0$, the set of conventional LMTO bases (1.5) is given. On the other hand, we can construct a short-range structure constant $\bar{S}_{R'L',RL}$, the so-called screened structure constant, by choosing appropriate $\bar{Q}_l$ values, because $\bar{S}_{R'L',RL}$ is given by the Dyson-type equation (1.10).

The matrices $\bar{o}$ and $\bar{h}$ are obtained from the condition that the envelope function $|\bar{K}\rangle^\infty$ is connected smoothly with $|\bar{\chi}\rangle^\infty$ on the muffin-tin surface. As a result, the basis function (1.6) is localized because the off-diagonal part of $\bar{h}$ is given by $\bar{S}$. The screened structure constant matrix $\bar{S}$ can be directly calculated from the bare structure constant matrix S in the real space by means of (1.10).

Using the basis set in (1.6), we can describe the Hamiltonian and the overlap matrices as follows:

$$\begin{aligned}
\bar{H} &\equiv {}^\infty\langle\bar{\chi}| - \Delta + v|\bar{\chi}\rangle^\infty \\
&= (I + \bar{h}\bar{o})\bar{h} + (I + \bar{h}\bar{o})E_\nu(I + \bar{o}\bar{h}) + \bar{h}E_\nu p\bar{h} , \tag{1.11}
\end{aligned}$$

$$\bar{O} \equiv {}^\infty\langle\bar{\chi}|\bar{\chi}\rangle^\infty = (I + \bar{h}\bar{o})(I + \bar{o}\bar{h}) + \bar{h}p\bar{h} . \tag{1.12}$$

Now, we have a linear eigenvalue equation expressed in the form of $\bar{H}\varphi_n = E_n\bar{O}$, which determines the electronic structures. By orthonormalizing the basis, it can also be transformed to the following second-order Hamiltonian as a power series in $\bar{h}$ of the form

$$H = \bar{O}^{-1/2}\bar{H}\bar{O}^{-1/2} \approx E_\nu + \bar{h} + \bar{h}\bar{o}\bar{h} . \tag{1.13}$$

Thus, the Hamiltonian is transformed into the two-center tight-binding form with the localized basis from first principles, and we obtain the linear secular determinant in the form (1.2). The above method is referred to as the tight-binding LMTO method.

Recursion Method

The recursion method [19, 20] is based on the Lanczos method which transforms a Hermitian matrix into a tridiagonal matrix by a unitary transformation. The new basis set $\{|u_i\rangle\}$ is generated by the recurrence formula,

$$\begin{aligned}
b_1|u_2\rangle &= H|u_1\rangle - a_1|u_1\rangle , \\
b_n|u_{n+1}\rangle &= H|u_n\rangle - a_n|u_n\rangle - b_{n-1}|u_{n-1}\rangle . \tag{1.14}
\end{aligned}$$

The starting vector $|u_1\rangle$ is set to a unit vector such that only the ith component has nonzero value $|u_1\rangle_i = 1$, where i denotes the orbital whose projected DOS $n_i(E)$ is calculated. The coefficients a_n and b_n are defined by

$$a_n = \langle u_n | \boldsymbol{H} | u_n \rangle , \qquad b_n = \langle u_{n+1} | \boldsymbol{H} | u_n \rangle , \qquad (1.15)$$

so that $|u_l\rangle$ with $l \leq n - 2$ does not appear in (1.14). Then the Hamiltonian $\boldsymbol{H}$ is transformed into the tridiagonal form $\boldsymbol{H}'$ with the new basis set $\{|u_i >\}$ after N th recursion:

$$\boldsymbol{H}' = \{\langle u_n | \boldsymbol{H} | u_m \rangle\} = \begin{pmatrix} a_1 \ b_1 & & & & \Large 0 \\ b_1 \ a_2 \ b_2 & & & \\ & b_2 \ a_3 & b_3 & & \\ & & \ddots & \ddots & \ddots & \\ & & & b_{N-2} \ a_{N-1} \ b_{N-1} \\ \Large 0 & & & b_{N-1} \ a_N \end{pmatrix} . \qquad (1.16)$$

The Green function $G'(z)$ for this Hamiltonian is given by

$$G'(z) = \frac{1}{z - \boldsymbol{H}'} . \qquad (1.17)$$

The diagonal element is obtained in a continued fraction form, viz.,

$$G_{ii}(z) = G'_{11}(z) = \cfrac{1}{z - a_1 - \cfrac{b_1^2}{z - a_2 - \cfrac{b_2^2}{z - a_3 - \cfrac{b_3^2}{\ddots \cfrac{}{z - a_N - T^{(N)}(E)}}}}} , \qquad (1.18)$$

where $T^{(N)}(E)$ is a terminator at N th order recursion. The square-root terminator with the asymptotic recursion coefficients determined by Beer and Pettifor's method is usually used. Then the projected DOS is given by

$$n_i(E) = -\frac{1}{\pi} \mathrm{Im} \lim_{\eta \to +0} G'_{11}(E + i\eta) . \qquad (1.19)$$

In practice, it is most important to determine the appropriate order of the recursion process. As can be seen from (1.18), high-order recursion coefficients are needed to obtain high-energy resolution in the DOS. On the other hand, a recursion process of too high an order results in non-physical oscillation in the DOS spectrum because of the finite-size effect of the model atomic structure. Therefore, the order of recursion should be determined by a compromise between the energy resolution and the size of the model atomic structure.

Forced Oscillator Method

The forced oscillator method (FOM) was developed by Yakubo and Nakayama [21] and applied to ab initio calculations combining with the TB LMTO method by Tanaka et al. [22]. In the FOM, the eigenvalue analysis is mapped to a lattice-oscillation problem. The eigenvalue density is extracted by applying a periodic forced oscillation with corresponding frequency and tracing the time development of the lattice oscillation. The method has the following advantages over the recursion method:

– the eigenvalue density is evaluated within a particular energy range,
– the energy resolution of the resultant electronic structure can be controlled explicitly by controlling the forced-oscillation time.

The method is explained briefly below.

We shift the zero energy of the eigenvalue problem (1.2) by an appropriate amount E_0, so that $E_n + E_0$ is positive, and then obtain the following eigenvalue equation by substituting $H_{RL,R'L'}$ and ω_i^2 for $H_{RL,R'L'} + E_0\delta_{RR',LL'}$ and $E_n + E_0$ in (1.2),

$$\sum_{J'} H_{J,J'}\varphi_n^{J'} = \omega_n^2\varphi_n^J , \tag{1.20}$$

where J denotes a set of $\{R, L\}$, and φ_n^J means the Jth element of the state vector φ_n. The above eigenvalue problem is equivalent to a lattice-oscillation problem. We apply the periodic external force $F_J \cos t$ to each particle J. Then the equation of motion for each particle is written as

$$M_J\ddot{u}_J(t) = -\sum_{J'} \sqrt{M_J}H_{J,J'}\sqrt{M_{J'}}u_{J'}(t) + F_J \cos \Omega t , \tag{1.21}$$

where M_J and $u_J(t)$ are the fictitious mass and the displacement of the Jth particle, respectively. The displacement $u_J(t)$ can be decomposed into a linear combination of normal modes $\{\varphi_n^J\}$ with time-dependent coefficients $Q_n(t)$, in the form $u_J(t) = \sum_n Q_n(t)\varphi_n^J/\sqrt{M_J}$. The total energy $\mathcal{E}(t)$ of this lattice system at time t is given by the external force F_J, and shows resonance behavior at the frequency $\Omega = \omega_n$. It is given by

$$\mathcal{E}(t) = \frac{1}{2}\sum_n \left[\sum_J \frac{F_J}{\sqrt{M_J}}\varphi_n^J\right]^2 \frac{\sin^2[(\Omega - \omega_n)t/2]}{(\Omega - \omega_n)^2}$$

$$\to \frac{\pi t}{4}\sum_n \left[\sum_J \frac{F_J}{\sqrt{M_J}}\varphi_n^J\right]^2 \delta(\Omega - \omega_n) . \tag{1.22}$$

By setting the external force as $F_J = 2\sqrt{M_J}\delta_{Jj}$, (1.22) reduces to

$$\lim_{t\to\infty} \mathcal{E}(t) = \pi t \sum_n |\varphi_n^j|^2\delta(\Omega - \omega_n) . \tag{1.23}$$

The partial density of states $n_j(E)$ projected onto a state j for the original electronic structure problem is defined by

$$n_j(E) = \sum_n |\varphi_n^j|^2 \delta(E - E_n) = \frac{1}{2\Omega} \sum_n |\varphi_n^j|^2 \delta(E - E_n) \,. \qquad (1.24)$$

Comparing (1.23) with (1.24), we obtain the projected DOS as

$$n_j(E) = \lim_{t \to \infty} \frac{\mathcal{E}(t)}{2\pi t \sqrt{E + E_0}} \,. \qquad (1.25)$$

The energy resolution ΔE of the DOS obtained is determined by how closely the right-hand side of (1.23) approaches a δ function at a finite time t. It is evaluated by the full width at half height of the peak around Ω on the right-hand side of (1.23), and is expressed as a function of the forced oscillation time t and the energy E in the form $\Delta E = 11.12\sqrt{E + E_0}/t$. Thus the energy resolution of the resultant DOS can be determined and controlled by the forced oscillation time t.

Particle Source Method

The particle source method (PSM) is a quantum mechanical version of the FOM, which was developed by Iitaka [23] and applied to ab initio calculations by Tanaka et al. [24, 25]. In the PSM, the Green function operating on a state $|j\rangle$ is evaluated by numerically solving the following time-dependent Schrödinger equation with a single frequency source term:

$$i\frac{d}{dt}|\tilde{j}; t\rangle = H|\tilde{j}; t\rangle + |j\rangle\theta(t)e^{-i(E+i\eta)t} \,, \qquad (1.26)$$

where η is an infinitesimally small value and $\theta(t)$ is the step function. The solution of this equation with the initial condition $|\tilde{j}; t = 0\rangle = 0$ becomes

$$|\tilde{j}; t\rangle = -i \int_0^t dt' e^{-iH(t-t')}|j\rangle e^{-i(E+i\eta)t'}$$

$$= \frac{1}{E + i\eta - H}\left(e^{-i(E+i\eta)t} - e^{-iHt}\right)|j\rangle \,. \qquad (1.27)$$

We can neglect the second term in the brackets in the second row if the condition $\eta t \gg 1$ is satisfied. Then the solution of (1.26) after a long time T can be approximated, in terms of the Green function and the ket $|j\rangle$, with the relative accuracy $\delta = e^{-\eta T}$, as follows:

$$|\tilde{j}; T\rangle \approx \frac{1}{E + i\eta - H}|j\rangle e^{-i(E+i\eta)T}$$

$$= G(E + i\eta)|j\rangle e^{-i(E+i\eta)T} \,. \qquad (1.28)$$

The Green function operating on the ket $|j\rangle$ can be obtained as

$$G(E + i\eta)|j\rangle = \lim_{T \to \infty} |\tilde{j}; T\rangle e^{i(E+i\eta)T} \,. \qquad (1.29)$$

The matrix element of the Green function between the states $\langle i|$ and $|j\rangle$ can also be obtained as

$$\langle i|G(E + i\eta)|j\rangle = \lim_{T \to \infty} \langle i|\tilde{j}; T\rangle e^{i(E+i\eta)T} . \tag{1.30}$$

In practice, (1.29) and (1.30) are evaluated after the finite evolution time T defined by

$$T = -\frac{\log \delta}{\eta} , \tag{1.31}$$

with the required relative accuracy δ.

Since we can choose any state as the initial state $|j\rangle$ in (1.26), we can also evaluate the matrix elements of a product including several Green functions and other operators, such as $\langle i|AG(E + i\eta)AG(E + i\eta)|j\rangle$, by choosing a new initial state as

$$|j'\rangle = AG(E + i\eta)|j\rangle \tag{1.32}$$

in (1.26), and repeating the procedure described above. This is a great advantage when evaluating the various physical quantities described by the Green function. Applications of this method to conductivities in amorphous systems will be introduced in Sect. 1.2.5.

1.2.3 Electronic Structure and Magnetism of Amorphous Alloys

General Aspects

The first ab initio calculations of amorphous materials were carried out by Fujiwara, by combining the TB LMTO method with the recursion method [16]. Figure 1.2 shows the calculated DOS of amorphous Fe. Although detailed structures are smeared out, we can recognize the two-peak structure in the projected DOS of d-states. Even in an amorphous phase, the local atomic structure consists of a few types of polyhedra, as discussed in Sect. 1.2.1, and the electronic structure also reflects the local atomic structures. As a whole, the DOS shape resembles that in a close-packed crystalline structure, such as fcc or hcp, rather than that in a crystalline bcc structure, because Fe atoms tend to form a closely packed tetrahedral structure in the amorphous phase.

A remarkable feature is that the Fermi level is a little away from the peak position of the DOS, and located on the low energy side of the peak. As a result, the Stoner criterion is not satisfied in the amorphous phase, in contrast to bcc Fe. From this fact, Fujiwara predicted that amorphous Fe does not show uniform ferromagnetism. This was confirmed by the discovery of a phase transition from ferromagnetism to a spin glass in the vicinity of pure Fe in several Fe-based amorphous alloys.

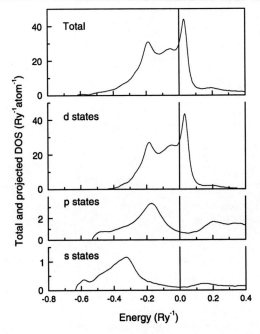

Fig. 1.2. Total and projected DOS for amorphous Fe [16]

The magnetism of amorphous transition metal alloys was first systematically investigated mainly in metal–metalloid alloys [26–28]. The data obtained showed that both the magnetization and the Curie temperature were uniformly lower than those of the crystalline counterparts when plotted as a function of average d-electron numbers. It was thus considered that the structural disorder which characterizes amorphous materials merely suppresses the magnetic properties for a long time.

However, the situation has changed since the discovery of spin-glass phases in amorphous transition metal alloys in the 1980s [29–34]. It was found that the ferromagnetism in Fe-rich amorphous alloys completely collapses beyond 90 at.% Fe, and that a new spin-glass (SG) phase appears. Furthermore, it was found that the Curie temperatures in Co-rich Y–Co amorphous alloys are enhanced as compared with those in their crystalline counterparts [35]. A similar enhancement of magnetism was observed in several Co-based rare earth–transition metal (RE–TM) amorphous alloys [36, 37].

These facts are related to the electronic structures of amorphous alloys. Kakehashi compared the DOS of Fe among amorphous [16] bcc and fcc [38] phases by plotting them with the same centers of gravity, and showed that the main peak of the DOS in the amorphous phase is located just between those in the bcc and fcc phases (see Fig. 1.3).

As a result, the Fermi level is away from the main peak and falls on the lower energy side in the amorphous phase. Consequently, ferromagnetism is

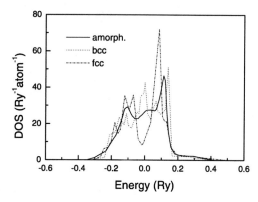

Fig. 1.3. Comparison of the total DOS's for Fe in amorphous [16], crystalline fcc [38], and bcc [38] phases

not expected according to Stoner's criterion, as mentioned above. Concerning the appearance of a spin glass phase, we have to resort to the finite temperature theory, which will be discussed in Sect. 1.2.4.

On the other hand, the Fermi level is expected to fall on the main peak position in amorphous Co on the basis of a rigid band model and its band-filling effect. This increases the exchange energy of amorphous Co and results in an enhancement of ferromagnetism. The enhancement of ferromagnetism observed in Co-based amorphous alloys at high Co concentrations (more than 80 at.% of Co) should be attributed to this mechanism. However, the enhancement observed in amorphous RE–Co alloys with 50 to 80 at.% of Co is due to the difference in local atomic structures between the amorphous and crystalline phases. This will be discussed in Sect. 1.2.3.

Amorphous Transition Metals

The spin-polarized DOS's of amorphous Co and Ni metals are shown in Fig. 1.4 [39]. The DOS's for these two metals are also quite similar, except that the exchange splitting is larger for amorphous Co than for amorphous Ni. Their shapes are well explained by the simple rigid band splitting of the paramagnetic DOS. In amorphous Co, the Fermi level is located between the main peaks of the majority and minority bands and this stabilizes the ferromagnetism of amorphous Co. On the other hand, it is located near the peak position of the minority band in amorphous Ni. This fact indicates that the ferromagnetism of amorphous Ni is unstable.

The calculated magnetic moment of amorphous Co is $1.63\mu_B$, which agrees well with the experimental values obtained from extrapolation to the pure amorphous Co limit in several Co-based amorphous alloys [40–42]. On the other hand, the calculated magnetic moment of amorphous Ni is $0.61\mu_B$,

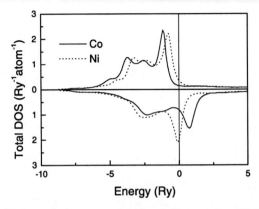

Fig. 1.4. Spin-polarized DOS for amorphous Co and Ni [39]

which also agrees well with the values extrapolated to the pure amorphous Ni limit in several Ni-based amorphous alloys [43–45].

The calculated magnetic moment of amorphous Ni is, however, very sensitive to the volume density compared with amorphous Co, and local magnetic moments in amorphous Ni are expected to fluctuate more widely than in amorphous Co, because of fluctuations in local volume densities.

Metal–Metalloid Alloys

The magnetic properties of metal–metalloid alloys have been extensively investigated in both their amorphous [46–48] and crystalline structures [49], and shown to be quite similar to each other in the two phases. The similarity is considered to originate from the similarity of the local atomic structure. It has been found that direct contact between metalloid atoms is prohibited from 0 at.% to about 20 at.% metalloids in metal–metalloid amorphous alloys [8]. Fujiwara et al. analyzed the local atomic structure of these materials using a relaxed DRPHS model [51] and showed that metalloid atoms were captured in the Bernal holes (see Sect. 1.2.1) formed by metal atoms. This local atomic structure is close to that of trigonal-prismatic crystalline Fe_3P or Fe_2P.

There have been several electronic structure calculations for amorphous Fe–B alloys [16,18,52,53]. Figures 1.5a and b compare the calculated DOS of amorphous $Fe_{80}B_{20}$ with that of crystalline Fe_2B [18]. The general features of their DOS's are almost the same, except for the fine structures observed in the crystalline phase. This is because the local atomic structures are quite similar to each other in these two alloys, as mentioned above. The calculated average magnetic moment per Fe atom is $2.1\mu_B$ in the amorphous phase, which agrees well with experimental values [28,54].

A pure amorphous Fe is expected not to be ferromagnetic, as mentioned before, because the Fermi level falls on the low energy side of the main peak

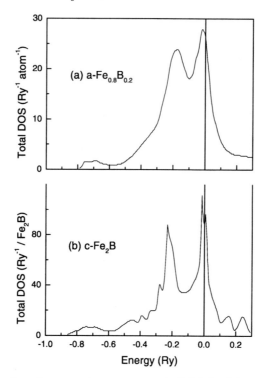

Fig. 1.5. Comparison between the paramagnetic DOS for (**a**) amorphous $Fe_{80}B_{20}$ and (**b**) crystalline Fe_2B [18]

in the DOS and Stoner's criterion is not satisfied. The ferromagnetism is, however, rapidly recovered by adding a few metalloid atoms, because the DOS at the Fermi level increases enough to satisfy Stoner's criterion. The increase in the DOS at the Fermi level is largely explained by two reasons [16]: one is the band narrowing of d-states due to the expansion of the Fe–Fe nearest-neighbor distance caused by adding metalloid atoms, and the other is the shift in the main peak position in the DOS towards the lower energy side due to the repulsion between the Fe d-states and the B p-states. These effects of introducing metalloid atoms are considered to be common to TM–metalloid alloys.

The amorphous $Co_{1-x}B_x$ system is a good example for the study of ferromagnetism, because it exhibits strong ferromagnetism at least in the range $0 \leq x \leq 0.3$ [45]. Tanaka et al. calculated the spin-polarized electronic structures of amorphous $Co_{1-x}B_x$ alloys ($x = 0.0$, 0.17, 0.23, and 0.32) using the TB LMTO recursion method [55].

Figure 1.6 shows the calculated total and projected DOS of the amorphous $Co_{83}B_{17}$ alloy. On the whole, the bonding nature is similar to that of Fe–B amorphous alloys. B sp-states split into bonding and antibonding states,

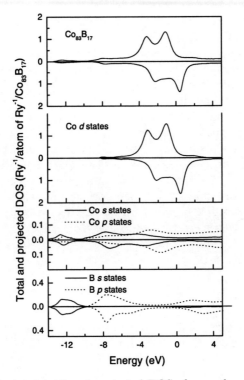

Fig. 1.6. Calculated total and projected DOS of amorphous $Co_{83}B_{17}$ [55]

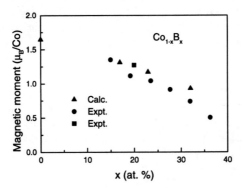

Fig. 1.7. Magnetic moment per Co atom for amorphous Co–B alloys obtained from electronic structure calculations (*filled triangles*) are compared with experimental ones (*filled circles* and *square*) [55]

and B p-states hybridize with the tails of Co d-states. The general shape of the DOS is well explained by a simple splitting of the rigid band in the paramagnetic state.

In Fig. 1.7, the calculated magnetic moments per Co atom are compared with experimental values [55]. Calculated values decrease with an increase in the B concentration and agree well with experimental values. The decrease in the magnetic moment was explained as resulting from the interatomic charge transfer from metalloid atoms to the minority TM d-state, and its band-filling effect [56]. On the other hand, Alben, Budnick, and Cargill suggested that the decrease in the magnetic moment originates from the loss of d-character because of hybridization with metalloid p-states [57]. The calculated results support the latter explanation. Although the interatomic charge transfer from B to Co increases with increased B concentration, the internal charge transfer from the majority Co d-states to the minority d-states decreases the magnetic moment of Co more significantly than does interatomic charge transfer. The internal charge transfer originates from the decrease in exchange splitting of Co d-states. The magnetic moment of Co is almost proportional to the exchange splitting of Co d-states [55] and is well explained by the generalized Stoner model [15].

RE–TM Alloys

The atomic structures of amorphous rare earth–transition metal (RE–TM) alloys have been extensively studied using the X-ray diffraction method. The obtained PDFs for several RE–TM amorphous alloys are quite similar to each other and can be well explained by a DRPHS model [7].

On the other hand, there are many stoichiometric crystalline phases such as RE–TM, RE–TM$_2$, RE–TM$_3$, RE–TM$_4$ and RE–TM$_7$, which are varieties of the Laves-phase RE–TM$_2$ alloy. The local atomic structures of RE–TM amorphous alloys are very different from those of their crystalline counterparts, in contrast to metal–metalloid systems. For example, the coordination numbers and nearest-neighbor distances in amorphous Gd$_{33}$Fe$_{67}$ are compared with those in the crystalline Laves phase in Table 1.1 [7]. Although the Fe–Fe and Fe–Gd atomic distances in the amorphous phase are close to those in the crystalline phase, the Gd–Gd distance in the amorphous phase is a little larger than in the crystalline phase. Furthermore, the coordination number of the Gd–Fe pair is much larger in the crystalline phase than in the

Table 1.1. Comparison of the nearest neighbor distances (r_{ij}) and coordination numbers (N_{ij}) in amorphous Gd$_{33}$Fe$_{67}$ and its crystalline counterpart [7]

	Amorphous		Crystalline	
	r_{ij} [Å]	N_{ij}	r_{ij} [Å]	N_{ij}
Gd–Gd	3.47	6 ± 1	3.20	4
Gd–Fe	3.04	6.5 ± 0.6	3.06	12
Fe–Fe	2.54	6.2 ± 0.5	2.61	6

amorphous phase. These differences in the local atomic structure cause the differences in electronic structures and therefore magnetic properties between the amorphous and crystalline phases.

In amorphous Gd–TM (TM = Fe, Co, Ni) alloys, Gd moments couple antiferromagnetically with the TM moments because of negative exchange coupling, which originates from the bonding characteristic between RE $5d$- and TM $3d$-states. The projected DOS's of GdFe$_2$ in the amorphous phase are compared with those in the crystalline Laves phase in Figs. 1.8a and b [58]. The DOS's in the amorphous and crystalline phases have some common features. The center of Gd d-states is located above the Fermi level, while that of Fe d-states is located below the Fermi level. The tails of Gd d-states hybridize with the Fe d-states and spread below the Fermi level. Since the energy level of the Fe d-states is closer to the Gd d-states in the up-spin band (the minority spin of Fe) than in the down-spin band, the Gd d-states hybridize with the Fe d-states more strongly in the up-spin band than in the down-spin band (i.e., $J_{RT} < 0$). Consequently, the DOS of the Gd d-states accumulates below the Fermi level in the up-spin band more than in the down-spin band. This polarizes the Gd d-states and makes Gd atoms couple with Fe atoms antiferromagnetically via the intra-atomic d–f exchange interaction. Due to the ferrimagnetism in Gd–Fe alloys, the projected DOS of Fe d-states in the majority spin part is very different from that in the minority spin part. A simple splitting of the rigid-band picture is no longer applicable to these materials.

The calculated magnetic moments in the amorphous phase are $2.0\mu_B$ and $7.2\mu_B$ for Fe and Gd, respectively. They agree well with experimental data [59]. The magnetic moment for the Gd site is larger than $7.0\mu_B$, because not only f-states but also d-states are polarized, as discussed above.

The projected DOS for Fe d-states in the crystalline Laves phase has a roughly two-peak structure with a deep valley between these two peaks, while that in the amorphous phase has a large peak followed by a small shoulder-like peak on the lower energy side, as shown in Fig. 1.8. This shows good agreement with the XPS spectrum observed by Güntherodt et al. [60]. Thus the projected DOS of Fe d-states in the amorphous phase is very different from that in its crystalline counterpart. This difference in the projected DOS results from the difference in the local atomic structure. The local atomic structure of the Gd$_{33}$Fe$_{67}$ amorphous alloy is very different from that of its crystalline counterparts, as shown in Table 1.1.

Figure 1.9 shows the calculated DOS of the amorphous Gd$_{33}$Co$_{67}$ alloy [61]. There are some common features between the DOS of amorphous Gd–Fe and Gd–Co alloys. The tails of Gd d-states hybridize strongly with the Co d-states and spread below the Fermi level. The Gd atom couples ferrimagnetically with the Co atom because of the difference in the strength of the hybridization in the up-spin and down-spin band, as in the case of the Gd–Fe amorphous alloy. As a result, Gd d-states are polarized as well as the

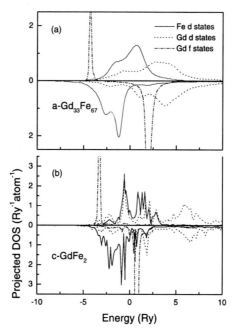

Fig. 1.8. Comparison of projected DOS for GdFe$_2$ in amorphous phase (**a**) and in crystalline Laves phase (**b**) [58]

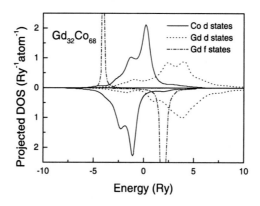

Fig. 1.9. Projected DOS of amorphous Gd$_{32}$Co$_{68}$ alloys [61]

f-states. The calculated DOS agrees well with the DOS observed optically using ultraviolet photoemission spectroscopy (UPS) [62], except for the position of f-states. In particular, Co d-states have a sharp peak at 1 eV below the Fermi level in both the calculated DOS and UPS data.

The calculated magnetic moments per Gd atom are $7.16\mu_B$ and $7.14\mu_B$ for Gd$_{18}$Co$_{82}$ and Gd$_{32}$Co$_{68}$, respectively. Fukamichi et al. [63] estimated the magnetic moment of Gd in Gd–Co amorphous alloys on the basis of the

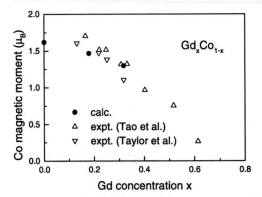

Fig. 1.10. Comparison of the magnetic moment per Co atom obtained by electronic structure calculations (*filled circles*) [61] and by experiment (*open triangles*) [36,37]

bulk magnetization and pressure derivative of the Curie temperature data. They found that the Gd moment is higher than $7.0\mu_B$ and that not only f-electrons but other electrons were also polarized. The calculated results agree with their observation.

Figure 1.10 compares the calculated magnetic moments per Co atom with experimental data. The magnetic moment per Co atom decreases with an increase in the Gd concentration. This decrease in the magnetic moment has been explained as a result of charge transfer from Gd atoms to minority Co d-states, and its band filling effect [36]. However, the numbers of charges are too small to explain the decrease in the magnetic moment per Co atom. The calculated results suggest that the decrease in the magnetic moment is caused by the reduction in the exchange splitting of the Co d-states, rather than by the charge transfer.

In amorphous Gd–Co alloys, it was found that magnetic properties such as the magnetic moment and magnetic ordering temperature are enhanced in a wide range of Co concentrations [36,37]. This enhancement is expected to result from the difference in local atomic structure between the amorphous and crystalline phases, in particular, the difference in the average coordination number around a Co atom.

A similar enhancement is observed in amorphous Y–Co alloys. Y–Co alloys are good reference materials for RE–Co alloys, because the electronic and atomic structures of RE–Co alloys are quite similar to those of Y–Co alloys, except that RE atoms have f-electrons. The ferromagnetism of amorphous Y–Co alloys is enhanced compared with their crystalline counterparts [35] in a wide range of Co concentrations. In particular, YCo_2 shows ferromagnetism in the amorphous phase, while it shows paramagnetism in the crystalline Laves phase.

In the crystalline YCo_2, the Fermi level is located a little away from a peak of the DOS and Stoner's criterion is not satisfied [64]. Ferromagnetism

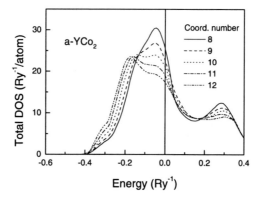

Fig. 1.11. Changes in the DOS for the amorphous $Co_{67}Y_{33}$ alloy as the coordination number around a Co atom varies from 8 to 12 [66]

is, however, realized in amorphous $Co_{67}Y_{33}$, because the Fermi level is located near a peak of the DOS and Stoner's criterion is satisfied [65]. This difference can be attributed to the difference in the local atomic structure between the crystalline and amorphous phases.

Kakehashi et al. systematically analyzed the effect of local atomic structure on the electronic structure [66]. They calculated the DOS of amorphous YCo_2 alloy by varying the average coordination numbers around a Co atom between 8 and 12 on the basis of a geometrical mean model and a Bethe-type approximation. Figure 1.11 shows the changes in the total DOS as a function of the coordination number.

The DOS at the Fermi level increases with decreasing coordination number and satisfies the Stoner criterion between 10 and 9. Note that the coordination number around a TM atom is 12 in the crystalline Laves phase, whereas it decreases to ≈ 9.5 in the amorphous counterpart [67]. In a binary dense random packing structure, the average coordination number around the smaller atom should be less than 12, the number expected from the close-packed structure, because the larger atom occupies a larger solid angle around an atom than does the smaller atom. This difference in the local atomic structure results in the different magnetism of YCo_2 alloys in the amorphous and crystalline phases. Although chemical bonding characteristics are important in the local atomic structure of amorphous alloys, the effect of the difference between the atomic radii of Co and Y is more significant in this case.

1.2.4 Finite Temperature Theory

Recently, a finite-temperature theory for amorphous metallic magnetism has been proposed by Kakehashi [68], based on the functional integral method for thermal spin fluctuations and the distribution function method for local magnetic moments with structural disorder. In this section, we explain the theory briefly, and introduce some applications to ferromagnetic–spin glass phase transitions.

Framework of the Theory

First, a fictitious exchange field variable ξ_i is introduced. The thermal average of the local moment (LM) on site i is then given by a classical average of ξ_i as

$$\langle m_i \rangle = \langle \xi_i \rangle = \frac{\int \left[\prod_j d\xi_j\right] \xi_i e^{-\beta E(\xi)}}{\int \left[\prod_j d\xi_j\right] e^{-\beta E(\xi)}} , \qquad (1.33)$$

$$E(\xi) = \int d\omega f(\omega) \frac{D}{\pi} \text{Im tr} \left[\ln(L^{-1} - t)\right] - N w_i(\xi) + \frac{1}{4}\tilde{J}\xi_i^2 , \qquad (1.34)$$

where $f(\omega)$ is the Fermi distribution function and t denotes the transfer integral t_{ij}. The locator L is defined by

$$\left(L^{-1}\right)_{ij\sigma} = L_{i\sigma}^{-1}\delta_{ij} = \left(\omega + i\delta - \epsilon_i^0 - w_i(\xi) + \frac{1}{2}\tilde{J}\xi_i\sigma + \mu\right)\delta_{ij} . \qquad (1.35)$$

It consists of the atomic level $\epsilon_i^0 - \mu$ measured from the chemical potential, the charge potential $w_i(\xi)$ which ensures charge neutrality, and the exchange potential $\tilde{J}\xi_i\sigma$. From (1.33), the fictitious field variable ξ_i is interpreted as a flexible LM on site i. It fluctuates thermally according to the energy functional $E(\xi)$.

We introduce the effective medium $\mathcal{L}_\sigma(\omega + i\delta)$ and expand the deviation from the medium with respect to the sites up to the second order. By using the molecular-field approximation for the LM on the nearest-neighbor (NN) shell, (1.33) reduces to

$$\langle m_0 \rangle = \frac{\int d\xi\, \xi\, e^{-\beta \Psi(\xi)}}{\int d\xi\, e^{-\beta \Psi(\xi)}} , \qquad (1.36)$$

$$\Psi(\xi) = E_0(\xi) + \sum_{j=1}^{z} \Phi_{0j}^{(a)}(\xi) - \sum_{j=1}^{z} \Phi_{0j}^{(e)}(\xi)\frac{\langle m_j \rangle}{x_j} , \qquad (1.37)$$

where z is the number of atoms on the NN shell. The atomic and exchange pair energies $\Phi_{0j}^{(a)}(\xi)$ and $\Phi_{0j}^{(e)}(\xi)$ are defined as follows:

$$\begin{bmatrix} \Phi_{0j}^{(a)}(\xi) \\ \Phi_{0j}^{(e)}(\xi) \end{bmatrix} = \frac{1}{2} \sum_{\nu=\pm} \begin{bmatrix} 1 \\ -\nu \end{bmatrix} \Phi_{0j}(\xi, \nu x_j) \ . \tag{1.38}$$

The quantity x_j in (1.37) and (1.38) is an amplitude in the single-site approximation, which is defined by

$$x_j^2 = \frac{\int d\xi \, \xi^2 \, e^{-\beta E_j(\xi)}}{\int d\xi \, e^{-\beta E_j(\xi)}} \ . \tag{1.39}$$

The function $\Phi_{ij}(\xi_i, \xi_j)$ denotes the pair energy between sites i and j,

$$\Phi_{ij}(\xi_i, \xi_j) = \int d\omega f(\omega) \frac{D}{\pi} \text{Im} \sum_{\sigma} \ln \left[1 - F_{ij\sigma} F_{ji\sigma} \tilde{t}_{i\sigma}(\xi_i) \tilde{t}_{j\sigma}(\xi_j) \right] \ , \tag{1.40}$$

where $\tilde{t}_{i\sigma}(\xi_i)$ is the single-site t matrix defined by

$$\tilde{t}_{i\sigma}(\xi_i) = \frac{L_{i\sigma}^{-1} - \mathcal{L}_{\sigma}^{-1}}{1 + \left(L_{i\sigma}^{-1} - \mathcal{L}_{\sigma}^{-1} \right) F_{ii\sigma}} \ . \tag{1.41}$$

The t matrix describes the impurity scattering when the impurity potential $\epsilon_i^0 - \mu + w_i(\xi) - \tilde{J}\xi_i\sigma/2$ is embedded in the effective medium $\mathcal{L}_{\sigma}^{-1}$.

Equations (1.36) and (1.37) bring out the fact that a flexible central LM ξ directly feels the molecular fields from the LMs $\{\langle m_j \rangle\}$ on the NN shell [the third term in (1.37)] and indirectly feels the average molecular field from the LMs outside the NN shell via the spin-dependent effective medium $\mathcal{L}_{\sigma}^{-1}$ which appears in $E_0(\xi)$, $\Phi_{0j}^{(a)}(\xi)$, and $\Phi_{0j}^{(e)}(\xi)$.

The effective medium $\mathcal{L}_{\sigma}^{-1}$ is determined by the coherent potential approximation (CPA) equation:

$$\left[\langle \tilde{t}_{i\sigma}(\xi) \rangle \right]_{\text{s}} = 0 \ , \tag{1.42}$$

where $\langle \ \rangle$ ($[\ ]_{\text{s}}$) denotes the thermal (structural) average. The off-diagonal disorder between the central and NN sites via the transfer integral t_{ij} is taken into account directly, whilst that outside the NN shell is described by an effective self-energy S_σ. The coherent Green functions $F_{ij\sigma}$ in (1.40) and (1.41) are therefore given by

$$F_{00\sigma} = \left(\mathcal{L}_{\sigma}^{-1} + \sum_{j=1}^{z} \frac{t_{j0}^2}{\mathcal{L}_{\sigma}^{-1} - S_\sigma} \right)^{-1} \ , \qquad F_{j0\sigma} = \frac{t_{j0}}{\mathcal{L}_{\sigma}^{-1} - S_\sigma} F_{00\sigma} \ . \tag{1.43}$$

The effective medium S_σ is determined from the condition that the structural average of the central coherent Green function $F_{00\sigma}$ should be identical with the neighboring one:

$$F_\sigma = [F_{jj\sigma}]_s = \int \frac{[n(\epsilon)]_s \, d\epsilon}{\mathcal{L}_\sigma^{-1} - \epsilon} , \qquad (1.44)$$

where $[n(\epsilon)]_s$ is the averaged DOS. Note that it can be calculated using the ab initio electronic structure calculation technique introduced in Sect. 1.2.2.

Equation (1.36) shows that the central LM is determined by the random variables of surrounding LMs $\{\langle m_j \rangle\}$ and the square of the transfer integral $\{y_j = t_{ij}^2\}$. Introducing the distribution $g(m_j)$ for surrounding LMs and probability $p_s(y_i)$ for the square of the transfer integral, the following integral equation is obtained from (1.36), since it is identical with that for the surrounding LMs:

$$g(M) = \int \delta(M - \langle m_j \rangle) \prod_{i=1}^{z} [p_s(y_i) \, dy_i \, g(m_i) \, dm_i] . \qquad (1.45)$$

This is the so-called the distribution function method. Equations (1.42) and (1.44) reduce to

$$\int \langle \tilde{t}_{0\sigma} \rangle \prod_{i=1}^{z} [p_s(y_i) \, dy_i \, g(m_i) dm_i] = 0 , \qquad (1.46)$$

$$\int F_{00\sigma} \prod_{i=1}^{z} [p_s(y_i) \, dy_i] = \int \frac{[n(\epsilon)]_s \, d\epsilon}{\mathcal{L}_\sigma^{-1} - \epsilon} . \qquad (1.47)$$

Finally, the LM distribution $g(M)$, the effective mediums $\mathcal{L}_\sigma^{-1}$, and $\mathcal{S}_\sigma$ are determined by solving (1.45), (1.46), and (1.47) self-consistently. The average magnetization $[\langle m \rangle]_s$ and the spin-glass order parameter $[\langle m \rangle^2]_s$ are obtained from the distribution $g(M)$ as follows:

$$[\langle m \rangle]_s = \int M g(M) \, dM , \qquad [\langle m \rangle^2]_s = \int M^2 g(M) dM . \qquad (1.48)$$

In practical calculations, we replace the distributions on the right-hand side of (1.45) and (1.47) by the binomial distributions and employ the equations which are correct up to the second moment. This means that the local environment is described whether each surrounding atom is located near or away from the average interatomic distance by a certain distance $[(\delta R)^2]^{1/2}$. Therefore, the input parameters needed in this theory are the exchange coupling $\tilde{J}$, the averaged DOS $[n(\epsilon)]_s$, the atomic coordination number z, and the fluctuation in the NN distance $[(\delta R)^2]^{1/2}/[R]_s$. The latter is estimated from the position and width of the first peak in the pair distribution function given by experiment.

Now the theory is extended in two directions. One is the extension to binary amorphous systems making use of a geometrical mean model [69], which enable us to study various magnetic properties of $3d$–$3d$, $3d$–$4d$, and $3d$–$5d$ amorphous transition metal alloys. The other is extension to a vector spin model by including the degrees of freedom for transverse moments [70, 71]. Some important applications are introduced in the next section.

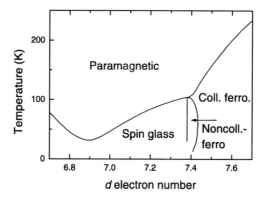

Fig. 1.12. Magnetic phase diagram of an amorphous transition metal as functions of temperature and d-electron number [71]

Spin Glass in Fe-Based Amorphous Alloys

Hiroyoshi and Fukami-chi [29] first suggested the existence of a spin glass in amorphous Fe_xZr_{1-x} alloys with $x \approx 0.9$. Saito and Nakagawa [30] and Coey et al. [31, 32] determined the magnetic phase diagram for amorphous Fe_xZr_{1-x} alloys up to 93 at.% Fe. They speculated that amorphous pure iron would be a spin glass with $T_g \approx 150$ K. Subsequently, Fukamichi et al. [33, 34] performed systematic investigations on amorphous Fe_xM_{1-x} (M = Y, Zr, La, Ce, and Lu) alloys. The important point is that beyond 90 at.% Fe, these alloys have the same spin-glass transition temperature irrespective of their second element M. This implies that the spin glasses for amorphous alloys with more than 90 at.% Fe are caused, not by the configurational disorder of Fe and M atoms, but rather by the structural disorder of amorphous pure Fe.

Kakehashi clarified the mechanism of the phase transition in detail on the basis of the finite temperature theory [68, 70, 71] introduced in the previous section. We shall present recent results obtained by the vector spin model [70, 71], which can describe the noncollinear ferromagnetic and spin-glass phases, although the essential picture for the appearance of a spin glass is not different from that obtained by the Ising spin model [68]. Figure 1.12 shows the calculated phase diagram of an amorphous transition metal as a function of d-electron number and temperature T. It shows that three phases appear at low temperatures depending on the d-electron numbers:

- spin-glass phase ($N \leq 7.38$),
- noncollinear ferromagnetic phase ($7.38 \leq N \leq 7.43$),
- collinear ferromagnetic phase ($7.43 \leq N$).

The spin-glass phase is further divided into two regions according to differences in the spin-glass formation mechanism.

In the region $6.9 \leq N \leq 7.2$, which corresponds to amorphous pure Fe, there exists nonlinear magnetic coupling between the moments of neighboring

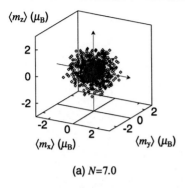

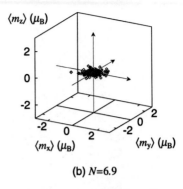

(a) N=7.0   (b) N=6.9

Fig. 1.13. Distributions of local moments: (a) at 30 K with a d-electron number of 7.0, and (b) at 80 K with a d-electron number of 7.42 [71]

transition metals. Local moments with large amplitudes couple ferromagnetically with neighboring moments, while those with small amplitudes couple antiferromagnetically with neighboring moments. Since the amplitudes of the local moments strongly depend on the surrounding environment, the sign of magnetic coupling changes with local environment and the spin-glass phase is realized.

In the region $7.2 \leq N \leq 7.38$, the local moments couple ferromagnetically with neighboring local moments. However, there exist long-range antiferromagnetic couplings, and these suppress long-range ferromagnetic order. Since this spin-glass phase is accompanied by ferromagnetic clusters, it is known as a cluster spin glass.

In both spin-glass phases, the distribution of local moments is spherical and indicates the appearance of an isotropic spin glass (see Fig. 1.13a). On the other hand, the transition from collinear ferromagnetism to noncollinear ferromagnetism is observed in the vicinity of the multicritical point on the magnetic phase diagram (see Fig. 1.13b). These are consistent with the experimental data of Mössbauer [72–74].

The finite temperature theory was extended to binary amorphous alloy systems by employing the geometrical mean method [69]. In Fig. 1.14 the calculated magnetic phase diagram for amorphous $Fe_x Zr_{1-x}$ is compared with the experimental one [34]. They show good agreement, except that transition temperatures are overestimated by a factor of 1.5–2 in the calculation. The overestimate is mainly due to the mean field approximation employed in the theory. In amorphous Zr–Fe, there coexist two effects causing the spin glass as described in amorphous pure transition metals: nonlinear magnetic coupling with the neighboring local moments and long-range antiferromagnetic coupling. However, the numerical results suggest that ferromagnetic couplings with neighboring local moments are rather strong as compared with antiferro-couplings. Therefore, long-range antiferromagnetic coupling seems a plausible explanation for the formation of a spin glass in Zr–Fe.

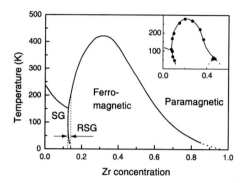

Fig. 1.14. Calculated magnetic phase diagram for amorphous Zr_xFe_{1-x} [69] compared with the experimental result (*inset*) [34]

The re-entrant spin glass behavior is also reproduced by the calculation near the ferromagnetic and spin-glass boundary, as shown in Fig. 1.14. This is mainly attributed to the thermal fluctuation of amplitudes of local moments. With an increase in temperature, the thermal spin fluctuation increases amplitudes of local moments. This makes the magnetic coupling with neighboring local moments more ferromagnetic, and unbalances the competition between short-range ferromagnetic and long-range antiferromagnetic interactions. As a result, a net magnetization remains at re-entrant spin-glass temperatures. These facts are consistent with the neutron scattering data, which suggests coexistence of the spin wave and spin glass in the re-entrant spin-glass region [75].

1.2.5 Transport Properties

DC Conductivities

Several calculations have been done to evaluate the dc conductivity of liquid and amorphous metals and alloys. Focusing on ab initio calculations, most of these are based on the Kubo formula and recursion method [76–78]. According to the Kubo formula, the dc conductivity at 0 K is given in the form

$$\sigma_{xx} = \frac{2\pi\hbar}{\Omega} \text{Tr}\left[\delta(E - H)j_x\delta(E - H)j_x\right] , \qquad (1.49)$$

where j_x is the current operator defined by $j_x = e[H, x]/i\hbar$ with the position operator x. Equation (1.49) can be rewritten as

$$\sigma_{xx} = \frac{e^2 n(E_F)D(E_F)}{\Omega} , \qquad (1.50)$$

where $n(E_F)$ is the DOS at the Fermi level and $D(E_F)$ is the diffusion constant defined by

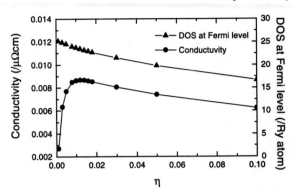

Fig. 1.15. Calculated dc conductivity and DOS at the Fermi level for amorphous Fe as functions of the energy resolution η [24]

$$D(E_F) = -\frac{\hbar}{e^2} \lim_{\eta \to 0} \mathrm{Im}\left[\langle E_F | j_x G(E_F + i\eta) j_x | E_F \rangle\right] , \qquad (1.51)$$

with eigenvector $|E_F\rangle$ at the Fermi level. The recursion method can evaluate this quantity, by choosing $j_x | E_F\rangle$ as the starting vector, and consequently the dc conductivity. In this method, however, we need to obtain the eigenvector $|E_F\rangle$ at the Fermi level using some other method, such as a filtering technique [79], because we cannot get reliable eigenvectors from the recursion method. Furthermore we have to repeat this process several times to take the average of the diffusion constant over eigenvectors around the Fermi level within a certain energy range.

Ballentine et al. calculated the dc conductivity of liquid La [76]. An interesting fact is that the contribution of d-states to the conductivity is dominant in liquid La. Although d-states are less mobile than s-states, d-states carry more electrons because they are superior in number on the Fermi surface. This is contrary to the usual concept that it is mainly sp-states that carry electrons.

The TB LMTO-PSM [24] introduced in Sect. 1.2.2 is a more sophisticated approach for calculating the dc conductivity. Using this method, we can directly evaluate the Kubo formula (1.49). The TB LMTO-PSM has the following advantages over the recursion method. It is not necessary to prepare the eigenvectors $|E_F\rangle$ at the Fermi level. The average of conductivity over the Fermi surface is taken automatically. There is no ambiguity, such as how to choose the terminator used in the recursion method. The only parameter required for the PSM is the energy resolution η, and we can evaluate both the conductivity and the DOS with the same algorithm.

Figure 1.15 shows the calculated conductivity and DOS of amorphous Fe as functions of the energy resolution η. The conductivity decreases rapidly below $\eta = 0.012$ with decreasing η, whereas the DOS increases monotonically

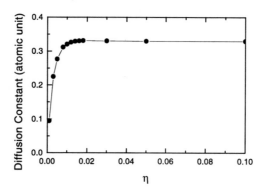

Fig. 1.16. Calculated diffusion constant for amorphous Fe as a function of the energy resolution η [24]

down to very small values of η. It is thus possible to get the extrapolated value of the DOS in the limit $\eta \to 0$.

The rapid decrease in conductivity comes from the discrete nature of the energy levels in a finite system. The value η represents the width of the delta-function-like imaginary part of the Green function. According to the Kubo formula, successive transitions must take place between the states within an energy resolution η in the vicinity of the Fermi level. Then it is necessary to choose η large enough to include a sufficient number of states.

On the other hand, both the conductivity and the DOS decrease gradually above $\eta = 0.012$ with increasing η. This fact reflects the degree of analytic continuation of the Green function onto the real axis in the complex energy plane. In this respect one should expect a better accuracy for smaller values of η. It is therefore a non-trivial task to find the most appropriate value of η. This problem is not specific to TB LMTO-PSM, but is in fact quite general.

The problem can be resolved by investigating the diffusion constant $D(E_F)$ evaluated from the conductivity σ_{xx} and DOS at the Fermi level $n(E_F)$ as

$$D(E_F) = \frac{\sigma_{xx}\Omega}{e^2 n(E_F)} \,. \tag{1.52}$$

It is plotted in Fig. 1.16.

A remarkable feature of the diffusion constant is that it remains almost constant throughout the whole resolution range, except below $\eta = 0.012$. This implies that the mobility of each electronic state is insensitive to the introduced resolution, so that we can safely extrapolate the constant value to $\eta = 0$ for this quantity. In order to reproduce the conductivity, we can make use of the DOS data extrapolated to $\eta = 0$.

The procedure gives $9.4 \times 10^{-3}/\mu\Omega$ cm for the conductivity. This is not far from the experimental value of $6.6 \times 10^{-3}/\mu\Omega$ cm [80], but overestimated by 42%. The contribution of d-states to the conductivity is estimated to be

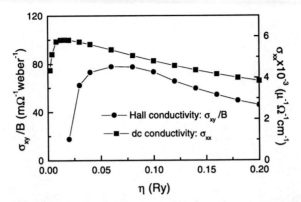

Fig. 1.17. Calculated dc conductivity and Hall conductivity for liquid Fe as functions of the energy resolution η [25]

about 80%. This result confirms the above argument given by Ballentine et al. [76].

Hall Conductivities

An interesting aspect of the Hall effect in disordered metals is the occurrence of the positive Hall coefficient. This was first found in liquid Fe and Co [81], and confirmed in a wide variety of materials [82]. A common feature of these systems is that they all include the transition or the rare-earth metals as constituent elements. Several explanations have been attempted for the appearance of a Hall coefficient with positive sign, but it remains an open question.

Itoh derived a rigorous expression for the Hall conductivity in the form [83]

$$\frac{\sigma_{xy}}{B} = -\frac{\pi e^3 \hbar^2}{c} \int \frac{\mathrm{d}E}{2\pi} \left(-\frac{\partial f}{\partial E} \right) \tag{1.53}$$
$$\times \mathrm{Im}\Big\{ \mathrm{Tr}\big[v_x G(E + i\eta) v_y \delta(E - H) v_x G(E + i\eta) v_y \delta(E - H) \big] \Big\},$$

where H and v are the one-electron Hamiltonian and the velocity operator without magnetic field, respectively. Tanaka et al. evaluated the Hall coefficient of liquid Fe using the above formula and TB LMTO-PSM [25].

Figure 1.17 shows the calculated dc and Hall conductivities, as functions of the energy resolution η, which is included in the formula (1.54). As can be seen from the figure, the Hall conductivity always shows a positive sign. Decreasing η from 0.20, both quantities increase monotonically at first, but the dc conductivity starts to decrease abruptly below $\eta = 0.01$. The Hall conductivity also follows a similar process of decrease, which starts at larger values of η, around $\eta = 0.06$.

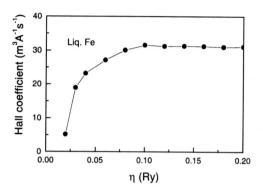

Fig. 1.18. Calculated Hall coefficient for liquid Fe as a function of the energy resolution η [25]

This abrupt decrease in the calculated conductivities with decreasing η occurs for the same reason as discussed in the calculation of dc conductivity. It comes from the discrete nature of the energy levels in a finite system. To evaluate conductivities, a sufficient number of states is necessary within the energy resolution η around the Fermi level, so that electrons can hop from state to state. The condition is more severe for Hall conductivity, because it is expressed in terms of four Green functions, instead of two in the case of the dc conductivity.

Although it is difficult to find the most reliable value of the Hall conductivity, we can make a reliable estimate of the Hall coefficient, which is described for weak magnetic fields by

$$R_{\mathrm{H}} = \frac{\sigma_{xy}/B}{\sigma_{xx}^{2}} \ . \tag{1.54}$$

As shown in Fig. 1.18, the Hall coefficient is insensitive to η in the range $\eta > 0.1$, and we can identify this value as the macroscopic Hall coefficient. It is evaluated to be $R_{\mathrm{H}} = 32.0 \times 10^{-11}\,\mathrm{m^3A^{-1}s^{-1}}$ and agrees quantitatively with the experimental value [81].

An important point is that the positive Hall coefficient can be explained by ordinary potential scattering alone, without recourse to another mechanism such as skew scattering or scattering due to magnetic disorder. It was also found that p–d hybridization plays an important role in the appearance of the positive Hall coefficient. Without p–d hybridization, the sign of the Hall conductivity becomes negative, and the projected density of p-states becomes close to a simple-metal DOS. When p–d hybridization is introduced, the p-states tend to separate into bonding and anti-bonding states and a shallow dip is created in-between. The Fermi level lies near the bottom of the dip.

A possible interpretation may be that 'holes' created in the bonding p band dominate the Hall conduction and become responsible for the positive contributions. From this viewpoint, there should be little difference between the disordered and crystalline states. Indeed many $3d$ transition metals (V, Cr, Fe and Co) show positive Hall coefficients in their crystalline states, and the same interpretation is expected to apply to these cases.

1.2.6 Conclusion

Although chemical short-range order is important, the local atomic structures of amorphous metals and alloys can be essentially explained by the dense random packing model, most of which consists of a few types of polyhedra. The electronic structure reflects this local atomic structure, and several properties which characterize amorphous phases can be explained by the difference in the local atomic structures between amorphous and crystalline phases. For the appearance of a spin-glass phase in the vicinity of pure amorphous Fe, structural disorder and the resultant competition between the ferromagnetic and anti-ferromagnetic couplings also play important roles. Recent progress in theory and computing enables us to evaluate conductivities of large disordered systems from first principles on the basis of the Kubo formula.

However, there are still many problems to solve. For example, many amorphous and liquid materials show a negative temperature coefficient of resistivity,

$$\text{TCR} = \frac{1}{\rho}\frac{d\rho}{dT} \, ,$$

but the mechanism is not clear for some of them. The first principles calculation does not reproduce the spin-glass phase of amorphous metals, even in the collinear spin configuration. Further theoretical progress is needed for the analysis of these problems from first principles.

1.3 Quasicrystals

1.3.1 Structures of Quasicrystals

Quasicrystals or quasilattices are characterized by the following facts [84,85]:

- the scattering intensity is a sum of densely distributed δ-functions,
- δ-function spots can be indexed by integers greater than the space dimension,
- the pattern of scattering spots has a rotational symmetry forbidden in crystals.

The quasicrystalline lattice is homogeneous with bond orientational order.

For nearly a decade after the discovery of quasicrystalline AlMn in 1984, many kinds of quasicrystal were found in a thermodynamically stable phase. Quasicrystals, both icosahedral and decagonal, are cluster compounds. Icosahedral quasicrystals can be classified into three different types. The first type

Table 1.2. Classification of quasicrystals

Classification		Materials
Icosahedral	SI	Al-Mn-Si, Al-Cr, Al-V-Si, Al-Pd-Ru, Al-Mn-Cu
MI-type	FCI	Al-Cu-TM(TM=Fe,Ru,Os), Al-Pd-TM(TM=Mn,Re)
Icosahedral	SI	Al-Li-Cu, Mg-Ga-Zn, Al-Mg-Pd, Al-Mg-Zn
TC-type	FCI	Mg-Li-Al, Mg-Zn-Y, Mg-Zn-RE(RE=Gd,Tb,Ho,Er)
Icosahedral Cd compounds	Binary	Cd-Yb, Cd-Ca
Decagonal		Al-Mn, Al-TM(TM=Fe,Pd,Os),Al-Co-Ni, Al-Cu-Co, Al-Pd-TM(TM=Fe,Ru,Os), Al-Cr-Si, Al-Ni-Rh
Octagonal		Cr-Ni-Si, V-Ni-Si, Mn-Si
Dodecagonal		Cr-Ni, V-Ni, V-Ni-Si, Ta-Te

is the Mackay icosahedron type (MI), such as i-AlMnSi and i-AlCuFe, whose unit of packing is an icosahedral cluster called the MI icosahedral cluster. Glue atoms are necessary to tie up such clusters [86]. MI-type quasicrystals contain transition metal elements and also Al atoms in many cases. The second type is the triacontahedron type (TC), such as i-AlCuLi and i-AlMgZn, whose fundamental units are triacontahedral atom clusters sharing several atoms with each other. The packing does not need any glue atoms in this case [86]. TC-type quasicrystals contain simple sp-metal elements instead of transition metal elements. The third type comprises the recently discovered binary $Cd_{5.7}Yb$ and $Cd_{17}Ca_3$ systems, which consist of icosahedral atom clusters and glue atoms. The Cd-based quasicrystals are the first stable binary systems. They are also unique in many respects [87]. A structural unit of decagonal quasicrystals is an atom column extending along one periodic direction, which holds five-fold rotational symmetry around the periodic axis. The decagonal and icosahedral quasicrystals are characterized by the golden ratio $\tau = (\sqrt{5}+1)/2 = 1.6180\ldots$. Several different types of quasicrystal are tabulated in Table 1.2.

1.3.2 Physical Properties and Electronic Structures

The above classification of icosahedral quasicrystals relates to the different ratios of atomic diameter d to quasilattice constant a_R,

$$\frac{d}{a_R} \approx 0.61 \text{ (MI)}, \quad \frac{d}{a_R} \approx 0.57 \text{ (TC)}, \quad \frac{d}{a_R} \approx 0.61\text{--}0.62 \text{ (Cd--R)}, \quad (1.55)$$

and different electron-to-atom ratios

$$\frac{e}{a} \approx 1.6\text{--}1.8 \text{ (MI)}, \quad \frac{e}{a} \approx 2.1\text{--}2.25 \text{ (TC)}, \quad \frac{e}{a} = 2.0 \text{ (Cd--R)}. \quad (1.56)$$

The actual e/a ratio is only allowed in a narrow range for stable icosahedral quasicrystals, and this fact implies that the stable quasicrystals are essentially stabilized by an electronic mechanism.

There are several crystalline compounds with close stoichiometry whose physical properties are very similar to those in quasicrystalline compounds. When the lattice periodicity of the crystalline compound is specified by a rational number approximation to the irrational number defining the non-crystalline rotational symmetry or the quasiperiodicity of quasicrystals, this crystalline compound is called a crystalline approximant. In most quasicrystals, various crystalline approximants are found. Atomic environments in crystalline approximants are locally similar to those in quasicrystals, because a crystalline approximant can be obtained by introducing phason defects or interchanging atomic sites in the quasicrystalline structure.

The electronic structures in one- and two-dimensional quasilattices have been thoroughly investigated. Their energy spectra and the spatial extent of their wavefunctions are quite anomalous [90–94]. The electronic properties in quasicrystals of perfect structural order are experimentally quite exotic [95]:

- The electronic resistivity is anomalously large at low temperatures, e.g., $\sim 1.0\,\Omega\,\mathrm{cm}$ in AlPdRe at 4.2 K, and samples with higher structural order show lower conductivity,
- the electric resistivity decreases with increasing temperature, e.g.,

$$\frac{\rho(4\,\mathrm{K})}{\rho(300\,\mathrm{K})} \approx 190 \quad \text{in AlPdRe} \,,$$

- the Hall coefficient and thermoelectric power are temperature-dependent,
- no Drude peak is observed in infrared spectra.

The temperature dependence of the conductivity may be expressed as

$$\sigma(T) = \sigma_0 + \Delta\sigma(T) \,, \tag{1.57}$$

where σ_0 is the residual conductivity at low temperatures and $\Delta\sigma(T)$ is the sensitively temperature dependent term. The sensitivity to structural ordering appears only in the first term σ_0. This equation contradicts the normal Boltzmann theory, and the relaxation time approximation may not be a good starting point for transport phenomena in quasicrystals.

The electronic band structures have been calculated in several idealized and realistic approximants of quasicrystals [96–102]. A pseudogap with a width of 0.5–1 eV is usually observed in the density of states (DOS) at the Fermi level, and the origin of the pseudogap is attributed to the Fermi surface–pseudo-Brillouin zone interaction. In addition, the wavefunctions are believed not to be spatially uniform and a large part of the weight is distributed on particular atom clusters.

Singular Properties of Electronic Structure
in Low-Dimensional Quasilattices

Before discussing the electronic structures in realistic quasicrystals and crystalline approximants, it is intuitively useful to know the electronic structures in a one-dimensional quasilattice. An infinite sequence consisting of two numbers τ and 1 is called the Fibonacci sequence when the sequence holds a self-generation rule $\tau \to \tau 1$ and $1 \to \tau$. This rule, starting from 1, generates a sequence $\tau 1 \tau \tau 1 \tau 1 \tau \tau 1 \tau 1 \tau \tau 1 \tau 1 \tau \ldots$, which can also be defined as a limit of a recursive sequence

$$S_{n+1} = S_n S_{n-1} \quad \text{with} \quad S_0 = \{1\} \quad \text{and} \quad S_1 = \{\tau\} . \tag{1.58}$$

This sequence is not periodic and the ratio of the total numbers of τ and 1 is, in the limit $n \to \infty$, the golden ratio $\tau = (\sqrt{5}+1)/2$. When a one-dimensional lattice has a structure equivalent to the Fibonacci sequence in an alignment of hopping integrals or atomic potentials, the system is called the Fibonacci lattice. The Fibonacci lattice is quite important and illustrates several key concepts of electronic structures in quasicrystals.

There are three types of energy spectrum in general: absolute continuous, point and singular continuous. The absolute continuous spectrum has a spectral measure $d\mu = n(E)dE$ and smooth DOS $n(E)$, as is generally the case for the DOS in crystals. The spectral measure of the point spectrum is a set of δ-functions defined on a countable number of points $\{E_i\}$, as is the case for the spectra of one- and two-dimensional random systems. The singular continuous spectrum covers the other cases, in which the integrated number of states below a certain energy increases continuously as a function of energy but is non-differentiable at any energy, e.g., the Cantor function, and the DOS is not well-defined [94].

At the same time, wavefunctions are also classified into three types: extended, localized and critical, corresponding to the absolute continuous, point, and singular continuous energy spectra, respectively. Extended wavefunctions are defined with asymptotic uniform amplitude $\int_{|r|<L} |\psi(r)|^2 dr \sim L^D$, where D is the spatial dimension. Localized wavefunctions are specified by square integrability, $\int_{|r|<\infty} |\psi(r)|^2 dr = \text{Const}$. The third type of wavefunction (critical) is neither extended nor localized and a typical example might be a power-law-type wavefunction $\psi(r) \sim |r|^{-\nu}$ with $\nu \leq D/2$ [93]. The energy spectrum of periodic systems is usually continuous. Randomness may bring about a point spectrum.

The unique character of the Fibonacci lattice is its energy spectrum. The tight-binding model, in which the hopping integrals take two values arranged in the Fibonacci sequence, is important for understanding the energy spectrum in quasiperiodic systems. The resultant energy spectrum of the tight-binding model has been mathematically proved to be singular continuous [90, 103].

A typical example of two-dimensional quasilattices is the Penrose lattice [104]. Electronic structures are analyzed by observing the behavior of the spectra and wavefunctions when changing the size of the periodic units of the crystalline approximants [93].

The calculated results bring out two important issues: one is that the DOS becomes less smooth and consists of sharper spikes with increasing system size. The spatial extent of wavefunctions can be studied by calculating the $2p$ norm of wavefunctions and the generalization of the participation ratio, together with multifractal analysis. Most eigenfunctions have power-law type in real space and cannot be normalized [93]. Conductance in the two-dimensional Penrose lattice has been calculated using the Landauer formula [105]. The resultant conductance has a power-law dependence on the linear size of the sample. The power-law dependence of the conductance is a direct result of the power-law decay of wavefunctions. Randomness smoothes the sharp spikes in the DOS and conductance channels open [101]. In other words, randomness mediates the hopping of electrons between certain power-law decaying eigenstates. This fact is consistent with the observed higher conductivity in samples of less ordered structures.

1.3.3 General Features of Electronic Structures in Realistic Quasicrystals

Structural Stabilization Mechanism 1: Fermi Surface–Pseudo-Brillouin Zone Interaction

In most quasicrystals or crystalline approximants, we observe an enhanced depression of the DOS at the Fermi energy, which is called a pseudogap [97]. The Bragg scattering in periodic systems modifies the DOS at Brillouin zone (BZ) edges $\pm K_0$ where the periodic potential splits degenerate band states $k \approx K_0$ and $-K_0$, and the resulting DOS has some depression and corresponding enhancement nearby.

The situation in quasicrystals may be similar to that in crystals. Bisecting planes tangent to principal scattering vectors correspond to principal diffraction spots and they play the role of principal scattering planes. These bisecting planes form a polyhedron of highly spherical shape called a pseudo-Brillouin zone (BZ), although a Brillouin zone cannot be defined in quasiperiodic systems.

If an effective wavevector of a wavepacket with the Fermi energy coincides with the principal scattering vector, electron wavefunctions interfere with each other, sometimes resulting in a depression of the DOS, i.e., we have pseudogap formation due to interference of the scattered electrons. If the Fermi energy is located nearby or just on the pseudogap, one can expect reduction of the total band energy due to the DOS being pushed downwards. This scenario constitutes the stabilization mechanism of the Fermi surface–pseudo-BZ interaction, or the Hume–Rothery mechanism [97, 106].

(a)   **(b)**

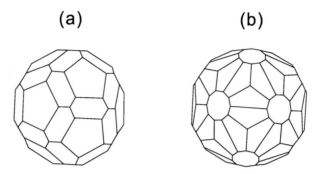

Fig. 1.19. The pseudo-Brillouin zone in quasicrystals. Pseudo-BZ constructed from (**a**) the [211111] and [221001] planes (MI type), and (**b**) the [222100] and [311111]/[222110] planes (TC type)

The pseudogap is more efficiently visible in quasicrystals than in crystalline approximants due to the pronounced sphericity of the pseudo-BZ in quasicrystals. The width of the pseudogap may be 0.5–1 eV. Several examples of the pseudo-BZ are shown in Fig. 1.19: the pseudo-BZ constructed from the [211111] and [221001] planes, appropriate to the MI type, and that from the [222100] and [311111]/[222110] planes, appropriate to the TC type.

The mechanism of the Fermi surface–pseudo-BZ interaction can be clearly seen in Table 1.3 [107], where the principal diffraction spots are indexed by 6-dimensional reciprocal lattice vectors and the corresponding critical valence per atom Z_c (electron-to-atom ratio). The coincidence between twice the radius of the Fermi surface and the radius of the principal diffraction spot $2k_F \approx K_p$ is actually satisfied in the TC-type icosahedral phase for strong diffraction spots (222100) and (311111)/(222110) with $Z_c \approx 2.1$–2.5 when we assume $Z_{Cu} = 1$. It is also satisfied in the MI-type icosahedral phase for scattering spots of (211111) and (221001) with $Z_c < 2.0$, if we adopt the concept of the negative valences of transition metal ions. The effects of strong sp–d hybridization may be attributed to the origin of the negative valences of transition metal ions [108].

Structural Stabilization Mechanism 2: sp–d Hybridization

It has been pointed out that pseudogaps may be widened and deepened by d-orbital resonance due to the virtual bound states in an icosahedral atom cluster [106, 109], although sp–d hybridization is not the principal origin of the pseudogap in MI- and TC-type quasicrystals.

The recently discovered binary quasicrystals Cd–Yb and Cd–Ca [87] may be the first whose stabilization mechanism is attributed to hybridization between the Cd sp-states and (Yb, Ca) d-states, which cause a shallow dip in the DOS near the Fermi level [127]. The origin of the shallow dip in the DOS is usually difficult to identify. However, in the present case, calculations for a

Table 1.3. Reciprocal lattice vectors and the corresponding critical valence per atom. The m_i are the components of the six-dimensional reciprocal lattice vectors [107]

m_1	m_2	m_3	m_4	m_5	m_6	Multiplicity of G	Critical valence Z_c TC type (AlCuLi)	MI type (AlMn)
2	1	1	1	1	1	12	1.28	1.5
2	2	1	0	0	1	30	1.49	1.75
2	2	2	1	0	0	60	2.17	2.55
3	1	1	1	1	1	72	2.42	2.84
3	2	2	1	0	1	60	3.91	4.59

hypothetical isostructural binary system Cd_6Mg in which the Mg ion has no d-state above the Fermi level do indeed give no shallow dip near the Fermi level. Therefore one can conclude that hybridization of the d-states near the Fermi level is essential for dip formation in Cd-based compounds.

Spikes in the Density of States

Another very characteristic structure in the calculated DOS is a dense set of very sharp spikes. This is inherent in the singular spectrum in one- and two-dimensional quasiperiodic systems. However, in three-dimensional quasiperiodic systems, it is still problematic [110] and there is no mathematical foundation for quasicrystals giving the singular spectrum. Furthermore, in almost all realistic quasicrystals, the chemical disorder is generic, and the mathematically idealized condition for the singular spectrum could not be satisfied.

Spatial Extent of Wavefunctions

The spatial extent of the wavefunctions in quasicrystals is crucial to understanding transport properties. One very intuitive example analyzed in some detail concerns wavefunctions in the decagonal quasicrystal AlCuCo [102]. The quasiperiodic plane of d-AlCuCo is constructed from large and small clusters. In a model, a large cluster contains 40 atoms on three shells and a smaller cluster contains 10 atoms on one shell with one atom at the center. There are no Co atoms in small clusters. An example of exact eigenstates near the Fermi energy is shown in Fig. 1.20. The wavefunction spreads favorably over larger clusters, and its participation ratio can be scaled by the number of atoms N as $P(\psi(E \approx E_F)) \propto N^{0.74}$. Of course, this is not the case for other eigenstates in different energy ranges. The power-law dependence of the participation ratio confirms the fact that wavefunctions near the Fermi energy only extend spatially over some specific alignment of atom clusters.

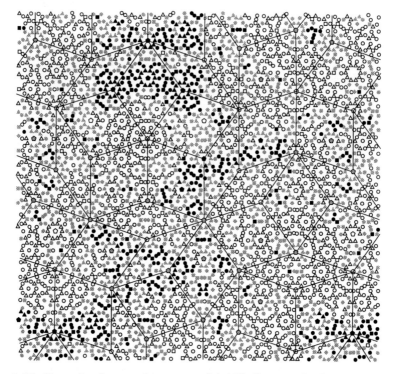

Fig. 1.20. Example of exact eigenstates of d-AlCuCo near the Fermi energy in a system of 2 728 atoms in a unit cell. *Symbols* show atomic positions of Al (○), Cu (△), and Co (□). Those atoms represented by *solid symbols* are most probable, with the total probability being 60%. Together, the *shaded* and *solid symbols* represent those atoms which have a total probability of 90%. The remaining atoms are represented by *open symbols*

1.3.4 Electronic Structure in Quasicrystals

Modification of the DOS in Model Icosahedral Al

A model calculation in a hypothetical quasicrystal of Al, using pseudopotentials, does indeed show a pronounced modification of the DOS from that of free electrons due to the scattering by the pseudo-Brillouin zone, as in Fig. 1.21 [111]. The DOS is largely modified, compared with fcc Al, by the icosahedral symmetry. This pronounced modification is associated with a large multiplicity in the peaks of the diffraction pattern. The Hume–Rothery stability mechanism causes this significant change in the DOS.

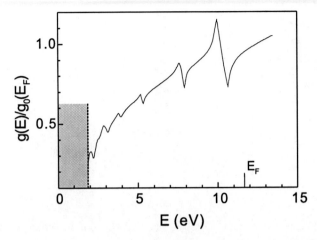

Fig. 1.21. Density of states in a hypothetical quasicrystal of Al [111]

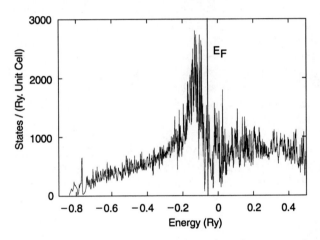

Fig. 1.22. Density of states in α-AlMn. The position of the Fermi level is shown by the *vertical line*. The main peak comes from the Mn d-state, and Mn $4s$-, Al $3s$- and $3p$-states spread over the lower and higher energy sides [98]

MI-Type Icosahedral Quasicrystals

AlMnSi

The electronic structure of idealized crystalline approximant α-AlMnSi was calculated [96,98] using the Elser–Henley model [112]. α-AlMnSi is the $(1/1)$ approximant, with 114 Al and 24 Mn, totalling 138 atoms per unit cell. Favorable positions of Si atoms are not known experimentally, and all Si atoms are substituted by Al atoms in the calculation.

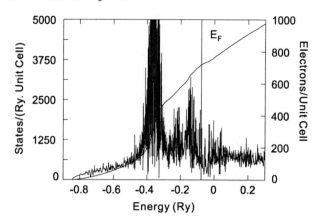

Fig. 1.23. Density of states in a hypothetical (1/1) approximant of AlCuFe [99]

The total DOS is shown in Fig. 1.22, where the DOS reveals a pseudogap at the Fermi energy with a width of 0.5 eV. The position of the pseudogap is near to those of strong scattering spots (211111) and (221001) and the pseudogap is therefore attributed to energy stabilization. The calculated Fermi level is located slightly below the minimum of the pseudogap. In the stable structure, some Al sites are substituted by Si atoms and the Fermi level climbs towards the minimum of the pseudogap. This is the stability enhancement mechanism of Si substitution.

The local and projected DOS show the standard resonance shape of the Mn d-states embedded in continuous spectra of Al and Mn s- and p-states. The contribution of Al d-states is appreciable. The Al d-states push down the Al p-states owing to the orthogonality within the same atomic sphere, and as a result also push down the Mn d-states. The resultant Fermi level shifts downwards by about 2 eV, but the bottoms of the conduction bands do not [98].

The center of a Mackay icosahedron has enough space for an Al atom but an Al atom cannot sit at this site, because not enough charge density could be contained in the local region of the center. Excess charge density flows outside the local region which corresponds to the region of the Fermi level.

AlCuFe

AlCuFe is a typical face-centered quasilattice and highly ordered samples can be prepared experimentally. The electronic structure was calculated [99] in a hypothetical model (1/1) approximant [113], consisting of $Al_{80}Cu_{32}Fe_{15}$ with 128 atoms per unit cell, and also in a realistic cubic (1/1) approximant structure [114] $(Al,Si)_{84}Cu_{36}Fe_{14}$ with 139 atoms per unit cell experimentally determined [115].

The total DOS of $Al_{84}Cu_{36}Fe_{14}$ is shown in Fig. 1.23 [114]. The Cu d peak is well separated from Fe d peaks in the DOS. A pseudogap is located in the

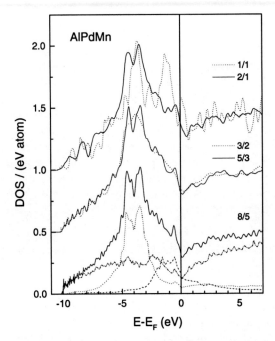

Fig. 1.24. Density of states of $(1/1)$ to $(8/5)$ approximants of AlPdMn. For the $(8/5)$ approximant, the local DOS is also shown: Al (*dashed line*), Pd (*dotted line*), Mn (*dot-dashed line*) [117]

vicinity of the Fermi energy with a width of 0.035 Ry, and the value of the DOS there is about 30% of that in pure crystalline Al. One can distinguish the local and projected DOS's for the same kind of atoms at inequivalent symmetry positions, and in particular one can see a large difference in the local DOS's at two different Fe sites. This suggests different covalent bonding and the role of glue atoms. Cu atoms, which have a d peak at lower energies than Fe atoms, have predominant stabilization effects with similar resonance effects in crystalline metallic ω-phase Al_7Cu_2Fe [108].

AlPdMn

Icosahedral AlPdMn is a typical stable face-centered quasilattice and exhibits very typical anomalous electronic transport properties. The lowest realistic crystalline approximant is the $(2/1)$ phase [116]. Crystalline approximants were constructed on the basis of the Katz–Gratius model and structural relaxation. Electronic structures were calculated from $(1/1)$ to $(8/5)$ approximants [117]. The total DOS's are shown in Fig. 1.24. The minimum of the pseudogap, located just above the Fermi energy in the $(1/1)$ approximant, shifts downwards with increasing unit cell size and, in the $(8/5)$ approximant, is located just at the Fermi energy. This gives a possible stabilization scenario in quasicrystals, more pronounced than in the crystalline approximants.

AlPdRe

AlPdRe is a unique icosahedral quasicrystal which exhibits peculiar be-havior in its transport properties [118–121]: $\rho \approx 1\,\Omega\,\mathrm{cm}$ at $4.2\,\mathrm{K}$ and $\rho(4\,\mathrm{K})/\rho(300\,\mathrm{K}) \approx 190$. Transport properties such as the conductivity, the Hall coefficient, and the magnetoconductivity, are very sensitive to the e/a ratio, which correlates strongly with the structural ordering [120]. Highly re-sistive samples of $e/a \approx 1.79$ show the metal–insulator transition at about $1\,\mathrm{K}$. The observed temperature dependence of the conductivity and mag-netoconductivity at very low temperatures could be explained only by the electron–electron interaction in the almost insulating regime, but could not be explained by the weak-localization theory. Photoconductivity was observed with complex relaxation processes [122].

TC-Type Icosahedral Quasicrystals

AlCuLi

The $(1/1)$ crystalline approximant of i-AlCuLi is Al_3CuLi_3, called the R-phase. This compound can be grown through thermodynamic processes but always contains intrinsic disorder because of the peritectic.

Electronic structures were considered in several structural models of the hypothetically ordered AlCuLi [97, 123]. The position of the pseudo-gap coincides with the principal peaks of the diffraction spots (222100) and (311111)/(222110). The pseudogap appears clearly in the model calculation, even in the Al–Li system without Cu atoms and the position and the width of the pseudogap do not alter [97]. The Fermi level shifts with changing atomic composition. This ensures stabilization by the pseudogap due to the Fermi surface–pseudo-BZ interaction. The Cu atoms do not contribute to formation of the pseudogap but supply additional electrons. The most appropriate con-tent of Cu atoms may be determined so as to adjust the Fermi level exactly to the bottom of the pseudogap. The height of the DOS at the bottom of the pseudogap is nearly $1/3$ of the free-electron parabola and is consistent with the observed reduction in the coefficient of the linear temperature-dependent term of the specific heat.

AlMgZn

Electronic structures of crystalline approximants of the TC-type icosahedral quasicrystals AlMgZn were calculated using the Henley–Elser model [124]. Systematic changes in the structures of the $(1/1)$ crystalline approximants were carefully analyzed and atomic positions were determined experimen-tally by the Rietvelt analysis. On the basis of refined structural models, the electronic structure was calculated with various atomic compositions $Al_xMg_{40}Zn_{60-x}$ with $15 < x < 53$ [126]. The position and width of the

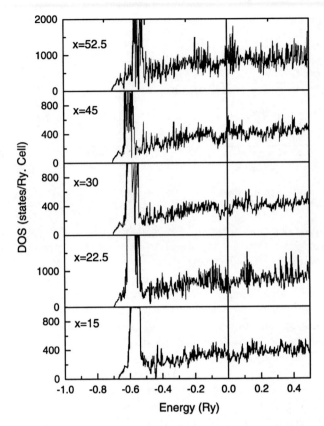

Fig. 1.25. Density of states of $(1/1)$ approximants of various atomic compositions of AlMgZn based on experimentally determined structural models [125]

pseudogap do not change appreciably with changing atomic compositions, but the Fermi energy gradually shifts across the valley of the pseudogap, as shown in Fig. 1.25. The observed XPS valence band spectra and also the concentration dependence of the experimentally derived electronic specific heat coefficient are consistent with the trends in the calculated spectra.

Binary Icosahedral Quasicrystals Cd–M (M = Yb, Ca)

Very recently, stable binary icosahedral quasicrystals $Cd_{5.7}Yb$ and $Cd_{17}Ca_3$ have been found [87–89]. Cubic crystalline phases, Cd_6M (M=Yb, Ca), exist in the composition near the quasicrystalline phase. The electron-to-atom ratio e/a is always 2.0 in Cd–Yb and Cd–Ca. The importance of the ratio of atomic radii is brought out. Isostructural cubic crystalline alloys are experimentally prepared for M = Sr, Y and most rare earth elements, but not in the quasicrystalline phase. The structure of Cd-based systems is also

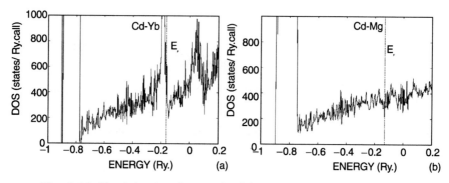

Fig. 1.26. Total density of states for (**a**) Cd_6Yb and (**b**) Cd_6Mg [127]

quite unique. The core of the icosahedral cluster is an atomic shell with non-icosahedral symmetry, that is a small cube of four Cd atoms placed at vertices with occupancy probability 0.5.

Figure 1.26 shows the total DOS of crystalline Cd_6Yb and Cd_6Mg [127]. It should be noticed that Cd–Mg is hypothetical. In Cd_6Yb, a shallow dip in the DOS is seen just below the Fermi energy between the occupied Yb $4f$ band and an unoccupied Yb $5d$ peak. The occupied states below the dip are predominantly made from the Cd $5p$-states, except for the narrow Yb $4f$ band. The strong hybridization of the Cd $5p$ and Yb $5d$ orbitals lowers the states of p symmetry near the Fermi level, thus leading to this dip. This statement can be confirmed by observing that there is no dip in the calculated DOS of Cd_6Mg, where magnesium has no d-state above the Fermi level. These results are contrasted with the cases of Al–Li–Cu and Zn–Mg–Y, where sp–d hybridization plays a rather minor role in pseudogap formation.

Decagonal Quasicrystals

Decagonal quasicrystals form a new class of anisotropic material with a crystalline axis (periodic direction) perpendicular to the quasicrystalline plane with five-fold rotational symmetry. This phase consists of two different classes, like the icosahedral phase: metastable systems (Al–Mn, Al–Fe, Al–Pd) and stable systems (Al–Cu–Co, Al–Ni–Co, Al–Pd–Mn). The local structure is similar to those in the crystalline approximant and the unit is a column cluster, i.e., a one-dimensional atom column with five-fold symmetry. The distance between layers along the periodic direction is about 2 Å. Decagonal AlCuCo and AlNiCo have two-layer periodicity with 4 Å period and others can be 8 Å (AlCo, AlNi), 12 Å (AlMn, AlPdMn), 16 Å (AlPd, AlCuFe), 24 Å and 36 Å, corresponding to 4, 6, 8, 12, and 18 layer structures.

AlCuCo

Several structural models have been proposed for d-AlCuCo [128,129]. Electronic structures were calculated on the basis of these models. There remains

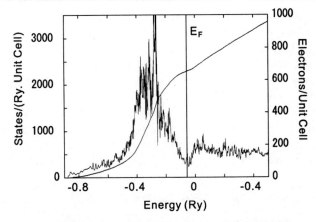

Fig. 1.27. Density of states of d-AlCuCo [100]

some ambiguity in atom positions in the models. Each atom column has two-layer periodicity.

The DOS based on the old model is shown in Fig. 1.27 [100], in the structure of which the Co sites are chosen so as to confirm a stable structure or show the pseudogap. The models contain two types of cluster. A large cluster contains 40 atoms on three shells with a vacancy at the center, 10 Cu on the inner shell, 10 Al on the middle shell and 20 Al and 10 Co on the outer shell. A small cluster contains one Al atom at the center and 5 Al and 5 Cu atoms on one shell. Two large clusters share 2 Al and 2 Co atoms with each other. There are no glue atoms. Although the pseudogap is sensitive in the structural models, the systematics of the $E(k)$ relation may be rather insensitive. The $E(k)$ curves are very flat with large effective masses within the quasiperiodic plane, while those along the periodic direction resemble the free-electron bands. Strong hybridization and large anisotropy of the energy band structure are observed. The spike structure is smoother in decagonal quasi crystals than in icosahedral quasicrystals, because of the one-dimensional periodicity.

In d-AlCuCo and d-AlNiCo, the ratio of electric conductivity along the periodic and quasiperiodic directions ranges from 4 to 12 [95,130]. The calculation based on the newer model enforcing a matching rule on clusters [131] reveals a non-apparent pseudogap. A recent, careful UPS experiment shows the existence of the pseudogap, both in d-AlCuCo and d-AlNiCo [132].

Experimental Observation of Electronic Structures in Quasicrystals

The pseudogap at the Fermi level is not a generic effect in quasicrystals, but is rather common in metallic compounds. Direct measurement of electronic structures has been carried out by XPS (X-ray photoelectron spectroscopy), UPS (ultraviolet photoelectron spectroscopy), IPES (inverse photoemission spectroscopy), SXES (soft X-ray emission spectroscopy), and SXAS (soft X-ray absorption spectroscopy). All these optical experiments show a pseudogap in the DOS [132, 133].

Among them, recent experiments using ultra-high resolution photoemission spectroscopy with a resolution of 5 meV is conclusive concerning the pseudogap in i-AlCuFe, i-AlCuRu, i-AlCuOs, i-AlPdMn, i-AlPdRe, i-ZnMgY, d-AlCuCo and d-AlNiCo [132]. Pseudogaps were also observed by tunneling spectroscopy [134], NMR [135], the temperature dependence of magnetic susceptibility, specific heat, etc. Recent NMR experiments reported the observation of a spike at the Fermi energy with a width of 20 meV [136].

1.3.5 Transport Properties in Quasicrystals

As already stated, the electronic properties in quasicrystals of perfect structural order are quite exotic. The electronic resistivity is anomalously large at low temperatures, and samples with higher structural order exhibit lower conductivity. The electric resistivity decreases with increasing temperature. No Drude peak is observed in infrared spectra.

The magnetoconductivity shows the crossover behavior $\Delta\sigma \sim -H^2$ (low H) to $\Delta\sigma \sim -\sqrt{H}$ (high H) below $T < 30\,\mathrm{K}$ in Al–Li–Cu. In Al–Cu–Fe, the magnetoconductivity behaves as $\Delta\sigma \sim -\sqrt{H}$ and its absolute value is reduced at higher temperatures [120, 137, 138].

Effects of Randomness

Chemical disorder is unavoidable in quasicrystals. Quasicrystals with positional randomness are highly conductive [139]. The conductance of the two-dimensional Penrose lattice with random phasons was investigated as a function of the system size, the random phason density and the Fermi energy [101]. Static randomness destroys the quasiperiodicity of the system and, as a consequence, the spikes in the fine structure of the DOS. Then the DOS is smoothed and electron transport channels open. This may be the interband transition due to elastic scattering by randomness, which makes electrons hop between weakly localized states and increases the conductivity in quasicrystals.

If the system length of two-dimensional model systems is increased, the electron conductivity shows additional crossover behavior from conductive to localizing, because in two-dimensional random systems, all states should be localized in a large enough system.

Boltzmann Theory

Transport properties were also analyzed on the basis of calculated band structures in crystalline approximants and the Boltzmann theory [140]. The calculated conductivity is generally very small if one assumes a reasonable value for the relaxation time. The conductivity fluctuates as a function of the Fermi energy and the fluctuation amplitude is the same order as the value of the conductivity itself. The small conductivity cannot be explained by the pseudogap alone, i.e., by the small carrier density, but small effective masses are very important in this analysis [98–100]. The energy difference between bands is 10–100 meV. Then the inelastic scattering at finite temperatures mediates the hopping of electrons in a weakly localized eigenstate to other empty eigenstates.

However, the relaxation time approximation cannot explain the observed behavior $\sigma(T) = \sigma_0 + \Delta\sigma(T)$. A unified picture of anomalous transport properties in quasicrystals cannot be constructed on the basis of band structure effects and the Boltzmann theory.

Scaling Behavior

The characteristic spatial extent of wavefunctions shows the power-law behavior of the participation ratio as a function of the system size. This can also be applied to the diffusion constant in quasicrystals [102].

The zero-temperature dc conductivity is expressed by the Kubo formula as

$$\sigma_{\alpha\beta} = \frac{2\pi e^2 \hbar}{V} \int dE \left(-\frac{df}{dE}\right) \sum_i \delta(E - E_i) D_i^{\alpha\beta}(E), \qquad (1.59)$$

$$D_i^{\alpha\beta}(E) = -\frac{1}{\pi} \lim_{\gamma \to 0+} \Im \left\langle i \left| \hat{v}_\alpha \frac{1}{E + i\gamma - \hat{H}} \hat{v}_\beta \right| i \right\rangle$$

$$\equiv \lim_{\gamma \to 0+} D_i^{\alpha\beta}(E : \gamma), \qquad (1.60)$$

where $\hat{H}$ is the Hamiltonian and $\hat{v}_\alpha = [\hat{x}_\alpha, \hat{H}]/i\hbar$ is the velocity operator. The function f is the Fermi–Dirac distribution function and V is the volume of the unit cell. This equation defines the diffusion constant $D_i^{\alpha\beta}(E)$ of the ith eigenstate, of energy E_i.

The value of γ in (1.60) should, in a strict sense, go to zero after the thermodynamical limit $V \to \infty$ has been taken. In the case of finite temperatures or randomness in the system, γ should remain finite. $D_i^{\alpha\alpha}(E : \gamma)$ was calculated for varying values of the parameter γ in several finite systems of volume V. When γ is unphysically small (much smaller than the averaged level interval δE proportional to $1/N$), then the behavior $D_i^{\alpha\alpha}(E : \gamma) \propto \gamma$ should be observed. For larger γ, D may be slowly varying. The crossover value γ_{cr} for two different types of behavior of D is the smallest limit of physically acceptable γ in a system of finite size.

This analysis was applied to d-AlCuCo and the scaling properties of the diffusion constant were discussed [102]. By enlarging the unit cell size, the crossover region is observed to shift gradually to the smaller γ side and smaller D values. We were able to fit the behavior of the crossover point to a curve with $D \sim \gamma_{\mathrm{cr}}^{0.25}$. The diffusion constant D may be written as $D \sim \langle r^2 \rangle \gamma_{\mathrm{cr}}$, where $\langle r^2 \rangle$ is the spatial extent of the wavefunction and $\hbar/\gamma_{\mathrm{cr}}$ is the mean free time. Assuming that the wavefunction is not strongly localized, we obtain $\langle r^2 \rangle \sim L^2$, where L is the effective relaxation length, proportional to the linear dimension of the system $\sqrt{N}$. We assume a scaling relation $\gamma_{\mathrm{cr}} \sim L^{-2\beta}$. Using the observed behavior $D \sim \gamma_{\mathrm{cr}}^{1-1/\beta} \approx \gamma_{\mathrm{cr}}^{0.25\cdots}$, one obtains an estimate $\beta \approx 1.33\ldots$ in d-AlCuCo. The diffusion constant of the finite system can then be written as $D \sim L^{-2(\beta-1)} \sim N^{-0.33\cdots}$, i.e., the diffusion constant should vanish in the thermodynamic limit at absolute zero temperature, $D \to 0$ ($N \to \infty$).

The value of the power-law index β is not universal and depends upon the energy region and the system itself, but the qualitative behavior may be quite universal. The diffusion constant D in an infinite system vanishes, but the decrease with increasing N is very slow. Actually, even if we were able to prepare a system with a γ_{cr} value of $10^{-6}\,\mathrm{Ry}$ ($\approx 1\,\mathrm{K}$), the diffusion constant would only become smaller by a factor of 10 in comparison to a system with $\gamma \approx 10^{-2}\,\mathrm{Ry}$.

The actual temperature dependence of the observed resistivity may be discussed in relation to the scaling behavior of the diffusion constant. Temperature causes incoherent electron scattering. γ then increases (for example, to γ_0) or the mean free path decreases. Then the whole system becomes equivalent to an array of block-perfect quasicrystals of length $L_0 \sim \gamma_0^{-1/(2\beta)}$. The bulk diffusion constant can be obtained by averaging over those of finite systems of length L_0.

Considering the scaling property and the experimental observation of the metal–insulator transition in AlPdRe at $1\,\mathrm{K}$, some additional mechanism may be required to achieve the metal–insulator transition at finite temperature, e.g., formation of a Coulomb gap due to electron–electron interactions.

Possibility of Metal–Insulator Transition

The conductivity at very low temperatures is observed to obey

$$\Delta\sigma(T) \propto T^{1/2} \qquad (1.61)$$

in several quasicrystals [95], and this was discussed in relation to the precursor of the metal–insulator transition in Al–Pd–$(\mathrm{Re}_{1-x}\mathrm{Mn}_x)$ [141]. It was also reported that the conductivity in Al–Pd–Re at very low temperatures 20–600 mK follows the temperature dependence of variable-range hopping [142]:

$$\sigma(T) = \sigma_0 \exp[-(T_0/T)^{1/4}]. \qquad (1.62)$$

The preceding arguments tell us that, although they are quite anomalous, the electron wavefunctions are not exponentially localized in quasicrystals. We know almost nothing about electron–electron correlation effects in quasicrystals, and the very narrow bands may create a situation where electrons have a strong correlation limit. Actually, observation of (1.61) implies the formation of a Coulomb gap. Further studies are therefore needed to clarify the possibility of a correlation gap for the narrow pseudogap observed in tunneling spectroscopy [143].

1.3.6 Conclusion

We have reviewed recent progress in the theoretical study of electronic structure in quasicrystals. Essential characteristics appear in the DOS, e.g., the pseudogap. We have summarized the cohesive mechanism in connection with pseudogap formation. The exact eigenstates in several crystalline approximants of decagonal quasicrystals are analyzed by finite-size scaling, and states near the Fermi energy show a clear power-law behavior in the system size. The amplitudes of wavefunctions are distributed on specific atom clusters, rather than uniformly over the whole system. However, these anomalous situations would not be enough to cause the metal–insulator transition. An additional mechanism, such as formation of a Coulomb gap due to electron–electron correlation, may give rise to the metal–insulator transition reported at very low temperatures.

1.4 Liquids

There are several different types of liquid which can be classified according to interparticle interactions. Liquids of rare gas atoms can be called atomic liquids, and systems consisting of independent nonpolar molecules may be called molecular liquids. A popular example of the latter is water (H_2O). Liquid metals are electron liquids in a positive charge background where the ionic distribution plays the role of a random potential moving much more slowly than the electron motion by a factor of a few tens or hundreds. Metal–insulator transitions have been observed in several liquid metals at very diluted densities, e.g., in cesium and in mercury. Liquid germanium and silicon, and presumably carbon, are metallic in liquid phase. Liquid tellurium and selenium are kinds of molecular liquids, forming large molecules with spiral chain structures.

X-ray and neutron experiments can give some information on atomic structures of liquids, such as the density and the pair distribution functions. However, detailed information can never be given by experiments and computer simulation is the only available resource.

1.4.1 First-Principles Molecular Dynamics Simulations

Car and Parrinello developed a new method of molecular dynamics simulation [5]. One usually calculates the electron structures in a fixed configuration of ionic positions $\{R_n\}$, and the electron charge density $n(r, \{R_n\})$ is evaluated for this configuration. The new idea is to treat the electron density $n(r)$ and ionic configuration $\{R_n\}$ as independent variables, and then find an extremum of the total energy in the extended configuration space $\left\{n(r), \{R_n\}\right\}$ or, more precisely, $\left\{\{\psi_i(r)\}, \{R_n\}\right\}$, where $\psi_i(r)$ is the electron wavefunction of the state i.

The total energy $E_{\mathrm{LDA}}(\{\psi_i(r)\}, \{R_n\}, \{\alpha_\nu\})$ includes the electron kinetic energy, electron–electron interactions, electron–ion interactions, and ion–ion interactions. The parameters α_ν are external constraints imposed on the system, such as the volume of the system, symmetry, the thermostat leading the system to a canonical distribution, and so on. The next step of the Car–Parrinello method is to introduce the Lagrangean

$$
L = \sum_i \frac{1}{2}\mu \int dr |\dot{\psi}_i(r)|^2 + \sum_n \frac{1}{2} M_n |\dot{R}_n|^2 + \sum_\alpha \frac{1}{2}\mu_\alpha \dot{\alpha}_\nu^2
$$
$$
- E_{\mathrm{LDA}}(\{\psi_i(r)\}, \{R_n\}, \{\alpha_\nu\}) , \tag{1.63}
$$

with the orthogonalization condition

$$
\int dr \psi_i^*(r)\psi_j(r) = \delta_{ij} .
$$

The dot denotes the time derivative, M_n is the ion mass, and μ and μ_α are arbitrary parameters. The first and third terms are the fictitious kinetic energies of the electrons and the external constraints. The actual electron kinetic energy is included in the total energy term E_{LDA}. The equation of motion for the variables can be derived from the Lagrangean (1.63) as

$$
\mu \ddot{\psi}_i(r, t) = -\frac{\delta E_{\mathrm{LDA}}}{\delta \psi_i^*(r, t)} + \sum_k \Lambda_{ik}\psi_k(r, t) , \tag{1.64}
$$

$$
M_n \ddot{R}_n = -\nabla E_{\mathrm{LDA}} , \tag{1.65}
$$

$$
\mu_\nu \ddot{\alpha}_\nu = -\frac{\delta E_{\mathrm{LDA}}}{\delta \alpha_\nu} . \tag{1.66}
$$

The Lagrange multipliers Λ_{ik} are introduced so as to satisfy the orthogonality constraint for the electron wavefunctions. The equation for the ionic motion is physical if the electron wavefunctions are on the Born–Oppenheimer surface, which can be specified by $\mathrm{Min}_{(\{\psi_i\}, \{\alpha_\nu\})} E_{\mathrm{LDA}}(\{\psi_i(r)\}, \{R_n\}, \{\alpha_\nu\})$. The equations for other variables $\{\psi_i\}$, $\{\alpha_\nu\}$ are fictitious, i.e., not real. If one can successfully reach an equilibrium state for an electronic configuration $\ddot{\psi}_i = 0$ and the external constraints $\ddot{\alpha}_\nu = 0$, then the wavefunctions satisfy the Kohn–Sham equations and are on the Born–Oppenheimer surface, and the external constraints are satisfied. The values of parameters μ and $\{\mu_\alpha\}$ should be chosen so that the equation of motion follows a trajectory not far from the Born–Oppenheimer surface. It is also very common now to solve $\ddot{\psi}_i = 0$ and $\ddot{a}_\nu = 0$ for (1.64) and (1.66), instead of solving the second-order differential equations [145, 146].

The first-principles pseudopotential is usually adopted, because the plane-wave base is very convenient for the calculations [3,4]. If we use a local basis set of electron wavefunctions and the centers of electron wavefunctions are chosen to coincide with the centers of moving ions, additional variational forces acting on ions and electron wavefunctions appear and create serious complexity. Furthermore, the transferability of the first-principles pseudopotentials (i.e., transferability to the same ion in a different local environment) is highly desirable when treating systems with moving local environment.

1.4.2 Electronic Structures in Liquid Metals

Structure of Simple Liquid Metals

The atom–atom interaction energy in simple liquid metals (sp-electron metals) is expressed as a sum of the two-body interaction which can be derived by the simple pseudopotential theory [147]. The resultant interatomic potentials can be expressed as

$$\phi(R_{ij}) \propto \frac{\cos(2k_{\mathrm{F}} R_{ij})}{(2k_{\mathrm{F}} R_{ij})^3} , \tag{1.67}$$

where k_{F} and R_{ij} are the Fermi wavenumber and the interatomic distance, respectively. This is a particular result of the metallic behavior of electrons in a weak potential field in simple metals. The resulting structure of liquid metals is essentially the dense random packing. The situation in liquid transition metals should be more complicated and the perturbation treatment for the pair potentials may fail.

The pair potentials in simple liquid metals is a function of parameters $x = 2k_{\mathrm{F}} R/2\pi$, $R_s = R_a Z^{-1/3}$ and R_{c}, where R, Z, R_a and R_{c} are the interatomic distance, the valence, the atomic radius (characterizing the density), and the radius of the empty core of the pseudopotential, respectively [148]. The pair potential can be reduced to the factorized form

$$\Phi(R) = \frac{Z^2}{R_a} \Phi_{\mathrm{red}}(x) , \tag{1.68}$$

where $\Phi_{\mathrm{red}}(x)$ is shown in Fig. 1.28. The oscillating behavior of the pair potential comes from the Friedel oscillation. The values of the phase and amplitude of the Friedel oscillations of the pair potential $\Phi_{\mathrm{red}}(x)$ are constant at fixed R_{c}/R_s. The first neighbor distance and further neighbor configurations may be determined by the position of the first minimum of the pair potentials and by the positions and depths of further deviations.

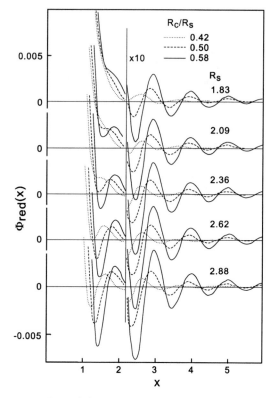

Fig. 1.28. Pair potential $\Phi_{\text{red}}(x)$ as a function of $x = 2k_{\text{F}}R/2\pi$ with the parameters $R_{\text{c}}/R_{\text{s}}$ and R_{s} [148]

The structures of liquid metals can be classified into three different types [149]. The first is the structure of groups I*a* and I*b* (Be, Mg, Al, Pb). The second type is the group of Zn, Cd, Hg and In. The last type is the structure of Ga, Si, Ge and Sn, whose structures deviate greatly from the random packing of hard spheres. These different types can be described by the pair potential and systematic variation of the valency Z or R_{s}, with constant $R_{\text{c}}/R_{\text{s}}$.

Electronic structures in Simple Liquid Metals

The systematic change in the electronic density of states (DOS) in simple liquid metals (*sp*-electron metals) has been investigated using the supercell model of 50–70 atoms in a unit cell generated by molecular dynamics simulations [150]. The agreement with experiment (electron energy loss spectroscopy) is fairly good. The resultant DOS's are shown in Fig. 1.29 and the results can be summarized as follows:

- For Li and Be, a strong reduction is found in the occupied band and the pronounced structure-induced minima in the DOS.

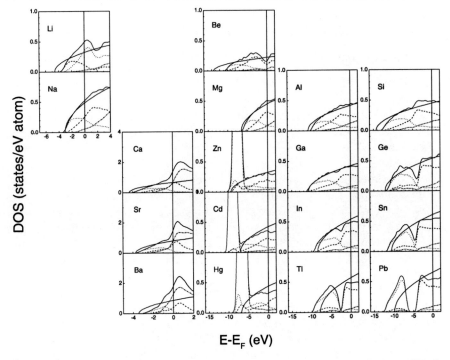

Fig. 1.29. Calculated density of states in simple liquid metals and group IV elements. *Continuous curve*: total density of states, *dotted curve*: s-states, *broken curve*: p-states, *chain curve*: d-states, *continuous parabola*: free-electron DOS. The *vertical line* shows the position of the Fermi level, chosen as the energy origin [150]

- The DOS's of Na, Mg, Al, and Si are close to those of the free-electron model.
- In alkaline-earth metals, s–d hybridization and the relativistic energy lowering of the s-states induce a significant deviation of the DOS from the free-electron model.
- In the group IIb elements (Zn, Cd, Hg), the narrow d band overlaps with the bottom of the broad sp band, and the DOS at the Fermi level is greatly reduced compared to the free-electron model.
- In In and Tl, a deep minimum of the DOS appears 2–3 eV below the Fermi level, similarly to the crystalline case. In Ga, the DOS seems to be very different from that in the crystalline case, whose DOS has a deep minimum at the Fermi level.

These results are obtained from calculations of the electronic structures, based on structural models derived by separated molecular dynamics simulation. The results should not be much altered if we use the Car–Parrinello molecular dynamics simulation, which simultaneously gives the atomic and electronic structures.

1.4.3 Electronic Structures
by First-Principles Molecular Dynamics

Group IV Elements: C, Si, Ge

In solid phase, Si and Ge are tetrahedrally bonded semiconductors of diamond structure with sp^3 hybridized orbitals. Tin has two solid phases: white tin (β-Sn) below 13.2°C and gray tin (α-Sn) above 13.2°C. Gray tin has the diamond structure and is a zero-gap semi-metal. White tin has the tetrahedral structure and is metallic. In heavier elements, the hybridization becomes weaker and the gap disappears.

Experimentally, liquid Si, Ge and even C are metallic and this fact is consistent with the calculation by the Car–Parrinello molecular dynamics simulation [151,152]. In the calculations, one can observe the pair distribution function, the velocity–velocity correlation functions, the distribution of bond angles, the atomic diffusion constant, the electronic DOS and a snapshot of the charge distribution, etc. The pair distribution function shows a wider distribution of the near-neighbor distances, which is characteristic of metallic systems. The number of nearest-neighbor atoms is about 6.4 in liquid Si, and about 2.9 in liquid C at 5000 K. This particularly low number of nearest-neighbor atoms in liquid C is due to the fact that the chains of carbon atoms are formed in the liquid phase. From the snapshot of the electron density distribution, one can observe that the bonds between atoms are formed and dissociated within a short time, something like 0.05 ps. Figure 1.30 shows a snapshot of the electron density distribution in liquid Si [151].

Other Liquid Metals

Liquid alkaline metals have been studied intensively [153]. The results by the Car–Parrinello simulation usually show good agreement with experiment, for example, in the structure and the diffusion coefficient, even near the triple point. Near the triple point the electron density spreads over the whole space, but at high temperatures the electron density tends to be localized due to a large spatial fluctuation in the atomic density.

In the case of heavy metals, the d-electron potential is deep and the usual pseudopotential cannot be used because it requires a large number of plane-wave basis functions. Liquid copper at a temperature of 1500 K was investigated using the Car–Parrinello molecular dynamics scheme based on the Vanderbilt ultrasoft pseudopotential [154]. The structural and dynamical properties are in excellent agreement with experimental data, such as the pair distribution function, and the diffusion constant.

Water

Water is a truly unique substance whose properties are determined by the hydrogen bonds and their network. The melting and boiling temperatures are

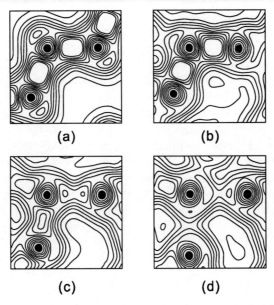

Fig. 1.30. Snapshot of the electron density distribution in liquid Si. *Filled circles* represent Si ions. The initial state (**a**) is the crystalline (110) surface. States (**b**)–(**d**) show the change in the electron density at 0.02 ps time intervals [151]

much higher than those of many other molecular liquids of similar molecular weight. The heat of evaporation and the heat capacity are very high.

It should be pointed out that the behavior of non-electrolyte solutions such as Ar, methane, and other nonpolar molecules with C–H base in water is anomalous. The solubility of these molecules is very low because of the hydrophobic base C–H. Hydrophobic solute molecules have a tendency to come together and attract each other in order to reduce the area touching water molecules (hydrophobic interaction). The ordered network of hydrogen bonds is broken in solutions, and the entropy increases. Therefore, the hydrophobic bond is stabilized at higher temperatures. Sometimes, water molecules form a more rigid hydrogen-bonded network around guest molecules (Ar, CH$_4$) with decreasing entropy. This is hydrophobic hydration. We should also mention that the hydrogen bond plays a very important role in the stereo structures of proteins.

In a water molecule, oxygen 2s and 2p orbitals form sp^3 hybridized orbitals. Then two electrons from two hydrogen atoms and two electrons from the oxygen atom occupy two bonding orbitals formed between the hydrogen 1s and the oxygen sp^3 orbital. The remaining four valence electrons of the oxygen atom occupy the unbounded sp^3 orbitals. The bond angle of H–O–H is 104.52°. The resultant symmetrical triad of point charges can be a model structure of the water molecule. This model ensures the existence of tetrahedrally directed hydrogen bonds. One model pair potential for a water molecule is based on a rigid four-point-charge model [155].

Recent investigations of water by the Car–Parrinello molecular dynamics simulation may be found in the literature [156]. The water monomer can be described rather nicely by the calculation. However, the electron correlation plays an important role in weak interactions such as hydrogen bonds and van der Waals interactions, because the correlation energy contributes a large part of the interaction energy.

An accurate description of hydrogen bonding in liquids requires an extension of the exchange functional such as the generalized gradient expansion. Moreover, ultrasoft pseudopotentials are crucial for describing the valence orbitals of oxygen atoms in a plane-wave expansion. The structural and dynamical properties of the liquid were found to be in good agreement with experiment. The lowest unoccupied molecular orbital of the liquid is the state occupied by a thermalized excess electron in the conduction band, which is a delocalized state distributed over interstitial space between the molecules with an admixture of the orbitals of the individual water molecules.

Ice is also a very interesting material comprising a hydrogen bond network. Several pressure-induced transitions of ice have also been studied using the Car–Parrinello molecular dynamics simulation. The phase transition is only described by using the gradient corrections to the local-density approximation, and cooperative dynamics is observed to be important.

Acknowledgements

We would like to thank Y. Kakehashi of the Hokkaido Institute of Technology for providing some figures prior to publication. We also thank Y. Shimizu of Shimane University for helping us to prepare figures.

References

1. P. Hohenberg and W. Kohn: Phys. Rev. B **136**, 864 (1964); W. Kohn and L.J. Sham, Phys. Rev. **140**, A1133 (1965); W. Kohn: Rev. Mod. Phys. **71**, 1253 (1999)
2. O.K. Andersen: Phys. Rev. B **12**, 3060 (1975)
3. D.R. Hamann, M. Schlüter, and C. Chiang: Phys. Rev. Lett. **43**, 1494 (1979); G.B. Bachelet, D.R. Hamann, and M. Schlüter: Phys. Rev. B **26**, 4199 (1982)
4. D. Vanderbilt: Phys. Rev. B **41**, 7892 (1990)
5. R. Car and M. Parrinello: Phys. Rev. Lett. **55**, 2471 (1985)
6. P.A. Lee and T.V. Ramakrishnan: Rev. Mod. Phys. **57**, 287 (1985)
7. G.S. Cargill: Solid State Phys. **30**, 227 (1975)
8. Y. Waseda: *The Structure of Non-Crystalline Materials* (McGraw-Hill, New York, 1980); R. Car and M. Parrinello: Phys. Rev. Lett. **25**, 2471 (1985)
9. J.D. Bernal: Nature **183**, 141 (1959); Nature **185**, 68 (1960)
10. J.D. Bernal: Proc. Roy. Soc. A **280**, 299 (1964)

11. C.H. Bennett: J. Appl. Phys. **43**, 272 (1972)
12. T. Ichikawa: Phys. Stat. Sol. (a) **29**, 293 (1975)
13. R. Yamamoto and M. Doyama: J. Phys. **F9**, 617 (1979)
14. C. Hausleitner and J. Hafner: Phys. Rev. B **45**, 128 (1992)
15. O.K. Andersen, O. Jepsen, and D. Glötzel: In *Highlights of Condensed Matter Theory*, ed. by F. Bassani, F. Fumi, and M.P. Tosi (North-Holland, New York, 1985)
16. T. Fujiwara: J. Non-Cryst. Solids **61–62**, 1039 (1984); O.K. Andersen: Phys. Rev. B **12**, 3060 (1975)
17. H.L. Skriver: *The LMTO Method* (Springer-Verlag, Berlin, 1984)
18. H.J. Nowak, O.K. Andersen, T. Fujiwara, O. Jepsen, and P. Vargas: Phys. Rev. B **44**, 3577 (1991)
19. R. Haydock, V. Heine, and M.J. Kelly: J. Phys. C **8**, 2591 (1975)
20. V. Heine: Solid State Phys. **35**, 1 (1980)
21. K. Yakubo, T. Nakayama, and J. Maris: J. Phys. Soc. Jpn. **60**, 3249 (1991)
22. H. Tanaka and T. Fujiwara: Phys. Rev. B **49**, 11440 (1994)
23. T. Iitaka: *Computing the Real-Time Green's Functions of Large Hamiltonian matrices*, High Performance Computing in RIKEN 1995 (ISSN-1342-3428), 241 (1996)
24. H. Tanaka: Phys. Rev. B **57**, 2168 (1998)
25. H. Tanaka and M. Itoh: Phys. Rev. Lett. **81**, 3727 (1998)
26. T. Mizoguchi, K. Yamaguchi, and H. Miyajima: In *Amorphous Magnetism*, ed. by H.O. Hooper and A.M. de Graaf (Plenum, New York, 1973); T. Mizoguchi: AIP Conf. Proc. **30**, 286 (1976)
27. R. Hasegawa, R.C. O'Handley, L. Tanner, R. Ray, and S. Kavesh: Appl. Phys. Lett. **29**, 219 (1976)
28. R.C. O'Handley, R. Hasegawa, R. Ray, and C.-P. Chou: Appl. Phys. Lett. **29**, 330 (1976); R.C. O'Handley, R. Hasegawa, R. Ray, and C.-P. Chou: J. Appl. Phys. **48**, 2095 (1976)
29. H. Hiroyoshi and K. Fukamichi: Phys. Lett. **85** A, 242 (1981); J. Appl. Phys. **53**, 2226 (1982)
30. N. Saito, H. Hiroyoshi, K. Fukamichi, and Y. Nakagawa: J. Phys. F **16**, 911 (1986)
31. J.M.D. Coey, D.H. Ryan, and R. Buder: Phys. Rev. Lett. **26**, 385 (1987)
32. J.M.D. Coey, E. Batalla, Z. Altonian, and J.O. Ström-Olsen: Phys. Rev. B **35**, 8630 (1987)
33. H. Wakabayashi, K. Fukamichi, H. Komatsu, T. Goto, and K. Kuroda: Proc. Int. Symposium on Physics of Magnetic Materials (World Scientific, Singapore, 1987) p. 342
34. K. Fukamichi, T. Goto, H. Komatsu, and H. Wakabayashi: Proc. 4th Int. Conf. on Phys. Magn. Mater. (Poland), ed. by W. Gorkowski, H.K. Lachowics, and H. Szymczak (World Scientific, Singapore, 1989) p. 354
35. K. Fukamichi, T. Goto, and U. Mizutani: IEEE Trans. on Magn. MAG-**23**, 3590 (1987)
36. L.J. Tao, S. Kirkpatrick, R.J. Gambino, and J.J. Cuomo: Solid State Commun. **13**, 1491 (1973)
37. R. Taylor and A. Gangulee: J. Appl. Phys. **47**, 4666 (1976)
38. V.L. Moruzzi, J.F. Janak, and A.R. Williams: *Calculated Electronic Properties of Metals* (Pergamon, New York, 1978)

39. H. Tanaka and S. Takayama: J. Phys.: Condens. Matter **4**, 8203 (1992)
40. A. Lienald and J.P. Rebouillat: J. Appl. Phys. **49**, 1680 (1978)
41. M.A. Fremy, D. Gignoux, A. Lienald: J. Magn. Magn. Mater. **44**, 263 (1984)
42. K.H.J. Buschow: Proc. Materials Research Society Conf. (Strasbourg), ed. by M. von Allmen (Les Editions de Physique, Les Ulis, 1984) p. 313
43. N. Heinman and N. Kazama: Phys. Rev. B **17**, 2215 (1978)
44. R.C. Taylor and A. Gangulee: J. Appl. Phys. **47**, 4666 (1976)
45. U. Mizutani, M. Hasegawa, K. Fukamichi, Y. Hattori, Y. Yamada, H. Tanaka, and S. Takayama: Phys. Rev. B **47**, 2678 (1993)
46. F.E. Luborsky: *Ferromagnetic Materials*, Vol. 1, ed. by K.H.J. Bushow and E.P. Wohlfarth (North-Holland, Amsterdam, 1980) p. 451
47. K. Moorjani and J.M.D. Coey: *Magnetic Glasses* (Elsevier, Amsterdam, 1984)
48. P. Hansen: in *Handbook of Magnetic Materials*, Vol. 6, ed. by K.H.J. Bushow (North-Holland, Amsterdam, 1991) p. 289
49. O. Beckmann and L. Lundgren: in *Handbook of Magnetic Materials*, Vol. 6, ed. by K.H.J. Buschow (North-Holland, Amsterdam 1991)
50. Y. Waseda: Prog. Mater. Sci. **26**, 1 (1981)
51. T. Fujiwara and Y. Ishii: J. Phys. F: Met. Phys. **10**, 1901 (1980)
52. S. Krompiewski, U. Krey, U. Krauss, and H. Ostermeier: J. Magn. Magn. Matter. **73**, 5 (1988)
53. W.Y. Ching and Y.-N. Xu: J. Appl. Phys. **70**, 6305 (1991)
54. C.L. Chien and K.M. Unruh: Nucl. Instrum. & Methods **199**, 193 (1982)
55. H. Tanaka, S. Takayama, M. Hasegawa, T. Fukunaga, U. Mizutani, A. Fujita, and K. Fukamichi: Phys. Rev. B **47**, 2671 (1993)
56. K. Yamaguchi and T. Mizoguchi: J. Phys. Soc. Jpn. **39**, 541 (1975)
57. R.A. Alben, J.I. Budnick, and G.S. Cargill: In *Metallic Glasses*, eds. J.J. Gilman and H.J. Leamy (American Society of Metals, Metals Park, Ohio, 1978)
58. H. Tanaka, S. Takayama, and T. Fujiwara: Phys. Rev. B **46**, 7390 (1992)
59. K.H.J. Buschow, M. Brouha, J.M.N. Biesterbos, and A.G. Dirks: Physica B **91**, 261 (1977)
60. G. Güntherodt and N.J. Shevchik: AIP Conf. Proc. No. 29, 174 (1976)
61. H. Tanaka and S. Takayama: J. Appl. Phys. **70**, 6577 (1991)
62. S.S. Jaswal, D.J. Sellmyer, M. Engelhardt, and Z. Zhao: Phys. Rev. B **35**, 996 (1991)
63. K. Fukamichi, K. Shirakawa, Y. Satoh, T. Masumoto, and T. Kaneko: J. Magn. Magn. Mater. **54–57**, 231 (1986)
64. S. Asano and S. Ishida: J. Phys. F: Met. Phys. **18**, 501 (1988)
65. J. Inoue and M. Shimizu: J. Phys. F: Met. Phys. **15**, 1525 (1985)
66. Y. Kakehashi, H. Tanaka, and M. Yu: Phys. Rev. B **47**, 7736 (1993)
67. M. Matsuura, H. Wakabayashi, T. Goto, H. Komatsu, and K. Fukamichi: J. Phys. Condens. Matter **1**, 2077 (1989)
68. Y. Kakehashi: Phys. Rev. B **40**, 11059 (1989); **41**, 9207 (1990); **42**, 10820 (1991)
69. M. Yu, Y. Kakehashi, and H. Tanaka: Phys. Rev. B **49**, 352 (1994)
70. T. Uchida and Y. Kakehashi: J. Appl. Phys. **81**, 3859 (1997); T. Uchida and Y. Kakehashi: Physica B **239**, 504 (1997)
71. T. Uchida and Y. Kakehashi: Phys. Rev. B **64**, 4402 (2001)
72. H. Wakabayashi, T. Goto, K. Fukamichi, H. Komatsu, S. Morimoto, and A. Ito: J. Phys. Soc. Jpn. **58**, 3383 (1989)

73. I. Vincze, D. Kaptás, T. Kemény, L.F. Kiss, and J. Balogh: Phys. Rev. Lett. **73**, 496 (1994)

74. H. Ren and D.H. Ryan: Phys. Rev. B **51**, 15885 (1995); K. Fukamichi, T. Goto, H. Komatsu, and H. Wakabayashi: In Proc. 4th Intern'l Conf. on the Physics of Magnetic Materials (Poland, 1988), ed. by W. Gorkowski, H.K. Lachowics, and H. Szymczak (World Scientific, Singapore, 1989) p. 354

75. J.A. Fernandez-Baca, J.J. Rhyne, G.E. Fish, M. Hennion, and B. Hennion: J. Appl. Phys. **67**, 5223 (1990)

76. L.E. Ballentine and J.E. Hammerberg: Phys. Rev. B **28**, 1107 (1983); L.J. Ballentine and M. Kolár: J. Phys. C: Solid State Phys. **19**, 981 (1986)

77. G.-L. Zhao, Y. He, and W.Y. Ching: Phys. Rev. B **42**, 10887 (1990)

78. S.K. Bose, O. Jepsen, and O.K. Andersen: Phys. Rev. B **48**, 4265 (1993); S.K. Bose, O. Jepsen, and O.K. Andersen: J. Phys.: Condens. Matter **6**, 2145 (1994)

79. B. Kramer and D. Weaire: J. Phys. C: Solid State Phys. **11**, L5 (1978)

80. J.G. Wright: Inst. Phys. Conf. Ser. **30**, 251 (1976)

81. H.-J. Güntherodt and H.U. Küniz: Phys. Condens. Mater. **16**, 117 (1973)

82. U. Mizutani: Prog. Mater. Sci. **28**, 97 (1983)

83. M. Itoh: Phys. Rev. B **45**, 4241 (1992)

84. D. Levine and P.J. Steinhardt: Phys. Rev. Lett. **53**, 2477 (1984)

85. P.J. Steinhardt and S. Ostlund: *The Physics of Quasicrystals* (World Scientific, Singapore, 1987)

86. C.L. Henley: In *Quasicrystals*, ed. by T. Fujiwara and T. Ogawa, Springer Series in Solid State Sciences 93 (Springer-Verlag, Heidelberg, 1990)

87. A.P. Tsai, J.Q. Guo, E. Abe, H. Takakura, and T.J. Sato: Nature **408**, 537 (2000)

88. J.Q. Guo, E. Abe, and A.P. Tsai: Phys. Rev. B **62**, 14605 (2000)

89. H. Takakura, J.Q. Guo, and A.P. Tsai: submitted to Philos. Mag. Lett. (2000)

90. M. Kohmoto, L.P. Kadanoff, and C. Tang: Phys. Rev. Lett. **50**, 1870 (1983); S. Ostlund, R. Pandit, D. Rand, H.J. Schellnhuber, and E.D. Siggia: Phys. Rev. Lett. **50**, 1873 (1983)

91. M. Kohmoto, B. Sutherland, and C. Tang: Phys. Rev. B **35**, 1025 (1987)

92. T. Fujiwara, M. Kohmoto, and T. Tokihiro: Phys. Rev. B **40**, 7413 (1989)

93. H. Tsunetsugu, T. Fujiwara, K. Ueda, and T. Tokihiro: J. Phys. Soc. Jpn. **55**, 1420 (1986); Phys. Rev. B **43**, 8879 (1991)

94. T. Fujiwara and H. Tsunetsugu, *Quasicrystals: The States of the Art*, ed. by D.P. DiVincenxo and P.J. Steinhardt (World Scientific, Singapore, 1991); D. Mayou: *Lectures on Quasicrystals*, ed. by F. Hippert and D. Gratias (Les Editions de Physique, Les Ulis, 1994); C. Sire: *Lectures on Quasicrystals*, ed. by F. Hippert and D. Gratias (Les Editions de Physique, Les Ulis, 1994)

95. K. Kimura and S. Takeuchi: *Quasicrystals: The States of the Art*, ed. by D.P. DiVincenxo and P.J. Steinhardt (World Scientific, Singapore, 1991); J.S. Poon: Adv. Phys. **41**, 303 (1992); C. Berger: *Lectures on Quasicrystals*, ed. by F. Hippert and D. Gratias (Les Editions de Physique, Les Ulis, 1994)

96. T. Fujiwara: Phys. Rev. B **40**, 942 (1989)

97. T. Fujiwara and T. Yokokawa: Phys. Rev. Lett. **63**, 333 (1991)

98. T. Fujiwara, G. Trambly de Laissardière, and S. Yamamoto: Mater. Sci. Eng. A **179–180**, 118 (1994)

99. G. Trambly de Laissardière and T. Fujiwara: Phys. Rev. B **50**, 5999 (1994)

100. G. Trambly de Laissardière and T. Fujiwara: Phys. Rev. B **50**, 9843 (1994)
101. S. Yamamoto and T. Fujiwara: Phys. Rev. B **51**, 8841 (1995)
102. T. Fujiwara, T. Mitsui, and S.Yamamoto: Phys. Rev. B **53**, R2910 (1996)
103. M. Kohmoto: Phys. Rev. B **34**, 5043 (1986); B. Sutherland and M. Kohmoto: Phys. Rev. B **36**, 5877 (1987); H. Hiramoto and M. Kohmoto: Int. J. Mod. Phys. B **6**, 281 (1992)
104. M. Gardner: Sci. Am. **236**, 110 (1977); N.G. de Bruijn: Indag. Math. Proc. Ser. A **84**, 27, 39, 53 (1981)
105. H. Tsunetsugu and K. Ueda: Phys. Rev. B **43**, 8892 (1991)
106. J. Friedel: Helv. Phys. Acta **61**, 538 (1988)
107. V.G. Vaks, V.V. Kamyshenko, and G.D. Samolyuk: Phys. Lett. A **132**, 131 (1988); A.E. Carlsson and R. Phillips: *Quasicrystals: The States of the Art,* ed. by D.P. DiVincenxo and P.J. Steinhardt (World Scientific, Singapore, 1991)
108. G. Trambly de Laissardière, N. Nguyen Manh, L. Magaud, J.P. Julien, F. Cyrot-Lackmann, and D. Mayou: Phys. Rev. B **52**, 7920 (1995)
109. G. Trambly de Laissardière and D. Mayou: Phys. Rev. B **55**, 1 (1997)
110. E.S. Zijlstra: Euro. Phys. Lett. **52**, 578 (2000)
111. A.P. Smith and N.W. Ashcroft: Phys. Rev. Lett. **59**, 1365 (1987)
112. V. Elser and C.L. Henley: Phys. Rev. Lett. **55**, 2883 (1985); P. Guyot and M. Audier: Phil. Mag. B **51**, L15 (1985)
113. E. Cokayne, R. Phillips, X.B. Kan, S.C. Moss, J.L. Robertson, T. Ishimasa, and M. Mori: J. Non-Cryst. Solids **153–154**, 140 (1993)
114. S. Roche and T. Fujiwara: unpublished work and paper presented at the Conference *Aperiodic 97*
115. H. Yamada, W. Iwakami, T. Yakeuchi, U. Mizutani, M. Tanaka, S. Yamaguchi, and T. Matsuda: Proc. of 6th International Conference on Quasicrystals, ed. by T. Fujiwara and S. Takeuchi (World Scientific, Singapore, 1997)
116. A. Waseda, H. Morioka, K. Kimura, and H. Ino: Phil. Mag. Lett. **65**, 25 (1992)
117. M. Krajčí, M. Windisch, J, Hafner, G. Kresse, and M. Mihalkovič: Phys. Rev. B **51**, 17355 (1995)
118. H. Akiyama, Y. Honda, T. Hashimoto, K. Edagawa, and S. Takeuchi: Jpn. J. Appl. Phys. **32**, L1003 (1993); F.S. Pierce, S.J. Poon, and Q. Guo: Science **261**, 737 (1993); Q. Guo and S.J. Poon: Phys. Rev. B **54**, 12793 (1996)
119. A.D. Bianchi, F. Bommeli, M.A. Chernikov, U. Gubler, L. Degiorgi, and H.R. Ott: Phys. Rev. B **55**, 5730 (1997); Q. Guo and S.J. Poon: to be published in Phys. Rev. B
120. H. Sawada, R. Tamura, K. Kimura, and H. Ino: Proc. of 6th International Conference on Quasicrystals, ed. by T. Fujiwara and S. Takeuchi (World Scientific, Singapore, 1997) p. 631
121. C. Gignoux, C. Berger, G. Fourcaudot, J.C. Grieco, and H. Rakoto: Euro. Phys. Lett. **39**, 171 (1997)
122. M. Takeda, R. Tamura, Y. Sakairi, and K. Kimura: Proc. of 6th International Conference on Quasicrystals, ed. by T. Fujiwara and S. Takeuchi (World Scientific, Singapore, 1997)
123. M. Windisch, M. Krajčí, and J. Hafner: J. Phys. Condens. Matter **6**, 6977 (1994)
124. J. Hafner and M. Krajčí: Phys. Rev. Lett. **66**, 333 (1991); Phys. Rev. B **47**, 11795 (1993)

125. U. Mizutani and T. Takeuchi: to be published in Phil. Mag. B
126. U. Mizutani, W. Iwakami, and T. Fujiwara: Proc. of 6th International Conference on Quasicrystals, ed. by T. Fujiwara and S. Takeuchi (World Scientific, Singapore, 1997)
127. Y. Ishii and T. Fujiwara: Phys. Rev. Lett. **87**, 206408 (2001)
128. S.E. Burkov: Phys. Rev. Lett. **67**, 614 (1991); J. Phys. I France **2**, 695 (1992)
129. S.E. Burkov: Phys. Rev. B **47**, 12325 (1993)
130. S. Martin, A.F. Hebard, A.R. Kortan, and F.A. Thiel: Phys. Rev. Lett. **67**, 719 (1991); W. Yun-ping and Z. Dian-lin: Phys. Rev. B **49**, 13204 (1994); W. Yun-ping, L. Li, and Z. Dian-lin: J. Non-Crys. Solids **153-154**, 361 (1993)
131. M. Krajčí, J. Hafner, and M. Windisch: Phys. Rev. B **55**, 843 (1997)
132. Z.M. Stadnik, D. Purdie, M. Garnier, Y. Baer, A.-P. Tsai, A. Inoue, K. Edagawa, S. Takeuchi, and K.H.J. Buschow: Phys. Rev. B **55**, 10938 (1997)
133. E. Belin and A. Traverse: J. Phys. Condens. Matt. **3**, 2157 (1991); E. Belin and D. Mayou: Phys. Scr. T **49**, 356 (1993); E. Belin, Z. Dankhazi, A. Sadoc, and J.-M. Dubois: J. Phys. Condens. Matt. **6**, 8771 (1994); M. Mori, S. Matsuo, T. Ishimasa, T. Matsuura, K. Kamiya, H. Inokuchi, and T. Matsukawa: J. Phys. Condens. Matt. **3**, 767 (1991); H. Matsubara, S. Ogawa, T. Kinoshita, K. Kishi, S. Takeuchi, K. Kimura, S. Suga: Jpn. J. Appl. Phys. **30**, L389 (1991)
134. T. Klein and O.G. Symko: Phys. Rev. Lett. **73**, 2248 (1994); T. Klein, O.G. Symko, D.N. Davydov, and A.G.M. Jansen: Phys. Rev. Lett. **74**, 3656 (1995)
135. F. Hippert, L. Kandel, Y. Calvayrac, and B. Dubost: Phys. Rev. Lett. **69**, 2086 (1992)
136. X.-P. Tang, W.A. Hill, S.K. Wonnell, S.J. Poon, and Y. Wu: Phys. Rev. Lett. **79**, 1070 (1997)
137. A. Sahnoune, J.O. Ström-Olsen, and A. Zaluska: Phys. Rev. B **46**, 10629 (1992)
138. T. Klein, H. Rakoto, C. Berger, G. Fourcaudot, and F. Cyrot-Lackmann: Phys. Rev. B **45**, 2046 (1992)
139. Y. Akahama, Y. Mori, M. Kobayashi, H. Kawamura, K. Kimura, and S. Takeuchi: J. Phys. Soc. Jpn. **58**, 2231 (1989)
140. T. Fujiwara, S. Yamamoto, and G. Trambly de Laissardière: Phys. Rev. Lett. **71**, 4166 (1993)
141. C.R. Lin, S.L. Chou, and S.T. Lin: J. Phys. Condens. Matter **8**, L725 (1996)
142. J. Delahaye, J.P. Brison, and C. Berger: Phys. Rev. Lett. **81**, 4204 (1998)
143. L. Guohong, H. Haifeng, W. Yunpin, L. Li, L. Shanlin, J. Xiunian, and Z. Dianlin: Phys. Rev. Lett. **82**, 1229 (1999); D.N. Davydov, D. Mayou, C. Berger, and A.G.M. Jansen: Int. J. Mod. Phys. B **12**, 503 (1998)
144. R. Car and M. Parrinello: Phys. Rev. Lett. **55**, 2471 (1985)
145. M.C. Payne, M.P. Teter, D.C. Allen, T.A. Arias, and J.D. Joannapoulos: Rev. Mod. Phys. **64**, 1045 (1992)
146. G. Kress and J. Furthmüller: Comp. Mat. Sci. **6**, 15 (1996)
147. V. Heine: Solid State Phys. **24**, 1 (1970); W.A. Harrison: Phys. Rev. A **136**, 1109 (1964)
148. J. Hafner and V. Heine: J. Phys. F **13**, 2479 (1983)
149. J. Hafner: *From Hamiltonians to Phase Diagrams* (New York, Springer-Verlag, 1987)
150. J.Hafner and W. Jank: J. Phys. Condens. Matter **2**, SA239 (1990)

151. I. Stich, R. Car, and M. Parrinello: Phys. Rev. Lett. **63**, 2240 (1989)
152. G. Galli, R.M. Martin, R. Car, and M. Parrinello: Phys. Rev. B **42**, 7470 (1990)
153. B.J.C. Cabral and J.L. Martin: Phys. Rev. B **51**, 872 (1995); F. Shimojo, Y. Zempo, K. Hoshino, and M. Watabe: Phys. Rev. B **52**, 9320 (1995)
154. A. Pasquarello, K. Laaronen, R. Car, C. Lee, and D. Vanderbilt: Phys. Rev. Lett. **69**, 1982 (1992)
155. F.H. Stillinger and A. Rahman: J. Chem. Phys. **60**, 1545 (1974)
156. K. Laasonen, M. Spik, and M. Parrinello: J. Chem. Phys. **99**, 9080 (1993); M. Spik, J. Hutter, and M. Parrinello: J. Chem. Phys. **105**, 1142 (1996); L.M. Ramaniah, M. Bernasconi, and M. Parrinello: J. Chem. Phys. **109**, 6839 (1998); C. Lee, D. Vanderbilt, K. Laasonen, R. Car, and M. Parrinello: Phys. Rev. Lett. **69**, 462 (1992)

2 Novel Application
of Anomalous Small-Angle X-ray Scattering
to Characterization of Condensed Matter

Y. Waseda, K. Sugiyama, and A.H. Shinohara

Small-angle X-ray scattering (hereafter referred to as SAXS) has been widely used for studying structural features of materials in various fields such as physics, chemistry, metallurgy, and biology, because SAXS data enable us to obtain important microstructural parameters such as particle volume and shape of so-called structural inhomogeneities with colloidal size [1–3]. Since SAXS signals are analyzed on the basis of coherent scattering due to electron density inhomogeneities in the sample, SAXS studies are usually made using radiation whose energy is far from an atomic absorption edge of constituent elements, where the atomic scattering factor for each element is almost proportional to the atomic number. Therefore, the conventional SAXS analysis produces a theoretical difficulty in detecting the structural inhomogeneities caused by the distribution of near-neighbor elements.

However, when the energy of the incident X-ray is close to an absorption edge of a constituent element, the atomic scattering factor is no longer expressed by the energy independent form. This is well-recognized as anomalous X-ray scattering [4–6]. This anomalous behavior with respect to the scattering factors constitutes a significant breakthrough for the analysis of structural inhomogeneities by introducing scattering contrast due to elemental inhomogeneities in the sample. The availability of intense, continuously tunable X-ray sources of synchrotron radiation promotes the use of anomalous small-angle X-ray scattering (anomalous SAXS), and major advances have been witnessed over the last few years.

The main purpose of this article is to review the anomalous SAXS technique and demonstrate its usefulness for obtaining structural information around a specific element, which could not be done by the conventional SAXS method.

2.1 Fundamentals

It has been experimentally recognized that intense continuous scattering appears in the neighborhood of the zero angle ($< 2°$), when certain structural inhomogeneities exist in the sample of interest. In samples with only one kind of atom, inhomogeneities are pure density defects, voids, grain boundaries, dislocations or surface structures. In the case of multi-component materials, a variety of compositional or topological inhomogeneities can exist in the form of fluctuations in the local concentration of the different atomic species. Since such inhomogeneities are frequently observed in the decomposition process of supersaturated solid solutions and in colloidal suspensions in dimensions from several tens to several hundreds of times the X-ray wavelengths, SAXS is widely used for structural studies of materials with compositional (chemical) and topological (density) inhomogeneities on the mesoscopic length scale.

Diffraction is produced by the interference of waves scattered by an object and every electron becomes the source of a scattered wave in the case of X-rays. Considering the enormous number of electrons and the fact that a single electron cannot be exactly localized, it will be convenient to introduce the concept of electron density $\rho(r)$ first. Then, the amplitude of the scattered X-rays is written as [1]

$$F = \iiint dV_k \rho(\boldsymbol{r}_k) e^{-i\boldsymbol{q}\cdot\boldsymbol{r}_k} , \qquad (2.1)$$

where $\boldsymbol{r}_k$ is the position of electrons and the wave vector $q = 4\pi \sin\theta/\lambda$ is given by the scattering angle 2θ and the wavelength λ. Mathematically speaking, the amplitude F of diffraction in a certain direction is the Fourier transform of the electron density distribution within the object V. The coherent scattering intensity $I(q)$ is the absolute square, derived using the complex conjugate F^*,

$$I(q) = FF^* = \iiint \iiint dV_1 dV_2 \rho(r_1)\rho(r_2) e^{-iq(r_1-r_2)} . \qquad (2.2)$$

This is a Fourier integral again, involving only the relative distance $r_1 - r_2$ for every pair of points. When the above integration is made by the positional correlation over all possible orientations, e.g., molecules in solution or precipitate particles in a randomly oriented polycrystalline system, the simplification $\langle e^{-iqr} \rangle = \sin(qr)/qr$ can be accepted and hence the coherent scattering intensity is further enhanced by introducing the relative distance r and its auto-correlation $\tilde{\rho}^2(r)$,

$$I(q) = \int 4\pi r^2 \tilde{\rho}^2(r) \frac{\sin qr}{qr} dr . \qquad (2.3)$$

The auto-correlation deviates from the initial value $V\overline{\rho^2}$ to the constant value $V\bar{\rho}^2$ in the large r region [1]. This behaviour is in accordance with the fact that a constant value throughout the total volume acts like a blank object and cannot make any contribution to the diffraction pattern, except for the extremely small angle region. It is convenient, therefore, to drop this background at the outset, and to use the electron density fluctuation $\eta = \rho - \bar{\rho}$ instead of the density ρ itself. Equation (2.3) now takes the form

$$I(q) = \int 4\pi r^2 \tilde{\eta}^2(r) \frac{\sin qr}{qr} dr . \qquad (2.4)$$

Let us now consider a dilute distribution of identical inhomogeneous particles of constant electron density ρ_P, embedded in a medium of matrix ρ_M. The density difference $\Delta\rho = \rho_P - \rho_M$ will be relevant for the X-ray diffraction intensity and can be well approximated by using an X-ray scattering factor together with number densities for the particle and the matrix.

The atomic scattering factor of an atom for X-rays is frequently assumed to be proportional to its atomic number Z. However, the atomic scattering factor for an atom should in principle be expressed in the form [4]

$$f(q,\varepsilon) = f^0(q) + f'(\varepsilon) + if''(\varepsilon) ,\qquad(2.5)$$

where ε is the energy of an incident X-ray, related to its wavelength by the equation $\lambda = hc/\varepsilon$. The symbols h and c are Planck's constant and the velocity of light, respectively.

The first term corresponding to the Fourier transform of the electron probability density $\rho(r)$ is proportional to Z. This term is considered to be independent of the energy of the incident X-ray. On the other hand, the second and third terms, known as the anomalous scattering factors, are significant only in the energy region near the absorption edge of a constituent element in a sample.

As an example, Fig. 2.1 shows the anomalous terms of Zn and Bi, calculated by Cromer and Liberman's relativistic method [5, 7]. The real term f' indicates a sharp negative peak at the absorption edge. Its width is typically 50 eV at half maximum. The variation of f' is typically 15–20% of the normal scattering factor f^0 at the K absorption edge. There appears to be a substantial change (over 50%) at the L absorption edge for most of the elements. This marked variation in f' is utilized in the AXS study. The imaginary term of f'' is positive and distinguished only on the higher energy side of the absorption edge. Its variation at the lower energy side of the absorption edge is known to be quite small. It should be mentioned that Cromer and Liberman's method does not predict any fine structure in the anomalous terms of the nearest edge and extended high energy regions. However, when the absorption measurement is carried out with a reasonably high degree of precision, a dispersion relation is available for determining both anomalous dispersion terms.

The imaginary part f'' is directly related to the linear absorption coefficient μ through the equation

$$f''(\varepsilon) = \frac{mc\varepsilon}{2he^2 N_a}\mu(\varepsilon) ,\qquad(2.6)$$

where m and e are the mass and charge of the electron, respectively, and N_a is the number of atoms per unit volume. The real part f' is given by the following Kramers–Krönig relation as a function of photon energy ε:

$$f'(\varepsilon) = \frac{2}{\pi}\int_{\varepsilon'=0}^{\infty}\frac{\varepsilon' f''(\varepsilon')}{\varepsilon'^2 - \varepsilon^2}\mathrm{d}\varepsilon' .\qquad(2.7)$$

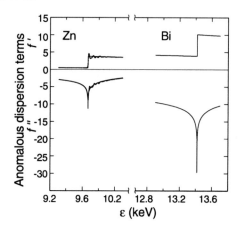

Fig. 2.1. Energy dependence of anomalous dispersion terms of Zn near the K absorption edge and Bi near the L_{III} absorption edge, computed using Cromer and Liberman's scheme [7]. The *dotted line* indicates the experimental terms of Zn in $ZnFe_2O_4$, calculated using the Kramers–Krönig relation [8]

An example of the experimental anomalous dispersion terms for Zn in the $ZnFe_2O_4$ structure is illustrated in Fig. 2.1 [8].

When the incidence of X-rays is tuned near the lower energy side of the absorption edge of a specific element A, the change in the real part f' of the anomalous dispersion terms of an element A produces significant intensity variations in the small-angle X-ray scattering. Hence, the scattering contrast detected at the absorption edge of an element A contains structural information with respect to the interesting element A alone. This significantly reduces the difficulty in understanding the environmental structure around an element A. A variety of analytical expressions were introduced for analyzing the SAXS curves near the angular origin, as will be exemplified by some selected examples. The fundamental idea is, however, to utilize the scattering contrast due to the anomalous dispersion terms in order to obtain environmental structural information around a specific element.

2.2 Experimental Procedure

Anomalous SAXS makes use of the variation of atomic scattering factors for X-ray energies in the vicinity of absorption edges. Synchrotron radiation is thus one of the most reliable and useful X-ray sources, because it provides intense and continuous X-rays. Usually, a monochromatic X-ray beam is tuned in the vicinity of the absorption edge of the relevant element. The monochromatic X-ray beam is defined in an incident beam with a typical cross section of about $1\,mm^2$ by an appropriate slit system, and its intensity is always monitored before reaching the sample by an ionization chamber or a monitor scintillator facing a polymer film. The sample is placed perpendicular to the incident X-ray beam and is separated by an evacuated chamber from the common linear position-sensitive detector or moving slit with a point detector. A number of experimental facilities for anomalous SAXS measurements have been developed recently and they include some further aspects given below [6, 9–11].

First Point

In ordinary synchrotron X-ray diffraction, the available energy range on the high energy side is limited by the characteristics of the radiation source. On the other hand, the high attenuation caused by the absorption of a sample, window materials and gas in the X-ray flight path gives a practical limit of about 4 keV at the lower energy side. In this context, the easily accessible elements for the anomalous SAXS have Z values heavier than 20 (Ca). Consequently, the most common elements encountered in polymer and colloid science (C, O, H and N) cannot be used as anomalous scatterers and this limits the usefulness of the technique to metals, electro-catalysts and charged colloids containing only relatively heavy elements. Nevertheless, recent careful design of a SAXS spectrometer with an ultrahigh vacuum flight path, in particular, allows us to challenge experimentally the phosphorus and sulfur K absorption edges [12, 13].

The inelastic processes at the higher energy side of the absorption edge, are known to produce a strong isotropic background of fluorescence, which inevitably disturbs precise intensity measurements. Consequently, anomalous SAXS experiments are usually performed at the lower energy side of the absorption edge of the interesting element. The values of f'' indicate relatively small values at the lower energy side of the absorption edge and no applicable energy dependence, so that the effect of the imaginary term on the measured scattering contrast can be safely neglected in the ordinary anomalous SAXS analysis.

In order to take advantage of the maximum changes in the atomic scattering factors in the vicinity of an absorption edge, an intense X-ray source with high energy resolution ($\Delta\varepsilon/\varepsilon = 10^{-4}$ and in any case $< 10^{-3}$) must be used. For this purpose, the ordinary monochromator system consists of two symmetrically cut perfect Si 111 or Si 311 crystals in a double-crystal configuration. The second crystal is frequently detuned out of maximum reflectivity by a fraction of its Darwin width so as to suppress the higher harmonic X-rays. It should be added that the existence of resonant Raman scattering and inevitable fluorescence due to the band pass of the monochromator limits the practical energy range to 5 eV or more below the absorption edge.

Any quantitative analysis of anomalous effects necessitates a knowledge of f' and f''. The theoretical values calculated by the Cromer–Liberman scheme are usually sufficient for energies on the lower energy side of the absorption edge [5, 7]. However, experimental determination of f' and f'' terms for an interesting element in the sample is absolutely necessary when X-rays in the vicinity or on the higher energy side of the absorption edge are used.

Second Point

In anomalous SAXS measurements, X-rays with energies between 5 and 20 keV are typically used as incident radiation. A suitable detector should respond well to this range of X-ray energies and indicate the availability of a high count rate. An anomalous SAXS measurement strongly prompts detectors with high energy resolution, because incident X-rays in the vicinity of the absorption edge of a target element sometimes partake in a strong incoherent interaction with the other constituent elements and produce a large background intensity of fluorescence. Two different modes of data collection are commonly used for the SAXS measurements: the sequential mode and the parallel mode.

In the sequential mode, a positioning device moves a receiving slit into a desired angular position and the radiation detector senses the scattered intensity at that position. Si and Ge solid-state detectors (SSD) are known to show the best energy resolution available at present. The energy resolution of a proportional counter (PC) and a scintillation counter (SC) are not as good as the solid-state detectors, but still sufficient for the SAXS measurement. PC has the advantage of a perfect response to soft X-rays with energies less than 9 keV and SC shows a uniform response to a wide range of X-ray energies.

The parallel mode of collecting SAXS data can be implemented using the position-sensitive proportional counter (PSPC). The application of the position-sensitive detector shortens the time of exposure by up to two orders of magnitude and accordingly lowers the radiation dose imposed on the sample. Samples with time-dependent properties can be investigated within the required short time. On the other hand, new problems arise when operating a PSPC, not encountered in the sequential mode: spatial resolution of the detector, deviation from uniform response at different positions, and count-rate capability. The detector efficiencies in position and energy are usually normalized by measuring the fluorescence intensity from a pure metal. The count-rate capacity is commonly estimated by including time-processing electronics and this inconvenience can be removed by selecting proper absorbing materials and adjusting dimensions. The distance between the sample and the position-sensitive detector can be adjusted so as to cover the required wave vector region.

Third Point

Characterization of the molecular weight and size for an inhomogeneous scatterer requires knowledge of scattering intensities in an absolute unit. Direct determination of the primary intensity is absolutely necessary for this purpose, although it is not usually easy due to the rapid succession of quanta in the primary beam. Several methods have been employed to overcome this difficulty. Luzzati's method [14] is one of the common techniques involving the determination of the absorption by several thin pure metallic foils, which are then stacked to yield sufficiently strong attenuation of the primary beam (by several orders of magnitude). Another, complementary technique is to use a secondary standard, such as a platelet of calibrated polyethylene (Lupolen 1811M of Badische Anilin & Sodafabrik). Note that the quotient of the absorption by the secondary standard and the sample under investigation should be well known and this involves determination of the absorption by the secondary standard. For these reasons, miniature ionization chambers, placed directly before and after the sample, are frequently used to measure the absorption and also to normalize the scattering data. The theoretical calculation of the absorption using the thickness and density of a sample coupled with the absorption coefficients as a function of the X-ray energy [5, 7] is a complementary method used for anomalous SAXS analysis.

It should be added that the necessity of obtaining absolute intensities at different energies with a precision better than a few percent requires more than just the standard corrections. It involves a precise subtraction of the background, including parasitic scattering from slits and window materials. In particular, the unfavorable remaining background intensity of fluorescence and Raman scattering is sometimes inevitable and makes a contribution to the measured scattering intensity. Such residual contributions can be successfully determined using the method known as Porod's asymptotic behavior in the large q region [3]. It should be mentioned that Porod's law only holds for idealized scattering particles with a smooth surface and step-like electron density difference from the matrix.

2.3 Selected Examples for Anomalous X-ray Scattering

The anomalous SAXS method has been successfully applied to structural investigation of various materials, in parallel with the use of an intense X-ray source of synchrotron radiation. It may be added that one of the main advantages in the experiments with the synchrotron radiation is to tune the energy of the incident X-rays to an appropriate energy range where anomalous scattering dominates. In this section, the potential of the anomalous SAXS method will be brought out by describing some selected examples obtained recently.

2.3.1 Particle Size Distribution

The determination of particle sizes and their distributions for metal-supported catalysts is of key importance in catalysis research, since the attainable number of active sites and electronic properties of catalysts are frequently given by their species as well as the shape and size of their loaded particles. Usually, metal-supported catalysts are three-phase systems of metal particles, the substrate support and its pores. Conventional SAXS analysis does not allow us to separate the scattering contribution of metal particles from the background scattering of the topological pore structure. This experimental inconvenience can be overcome by anomalous SAXS, which exploits the anomalous scattering feature that the scattering factors of metals vary at absorption edges, whereas the scattering factor of the non-metal substrate remains constant.

As an example, effective electrodes for electro-catalytic applications in fuel cells, for direct conversion of chemical energy into electrical energy, are commonly manufactured on the basis of electronically conductive porous carbon with noble metal catalysts such as Au, Pd and Pd–Au alloys in the form of small crystallites with diameters of around 1–5 nm and corresponding atomic fractions of about 10^{-3}. Conventional SAXS measurements cannot separate the structural information on the loaded metal particles within the carbon matrix. Anomalous SAXS measurements were therefore carried out in order to examine the size distribution of nanometer-sized catalyst particles (Haubold et al. [15, 16]).

Let us consider the small-angle scattering of particles of element A per sample volume V_s:

$$I_{\text{particle}}(q) = (n_A f_A - n_M f_M)(n_A f_A - n_M f_M)^* \frac{N}{V_s} \int_0^\infty P(r) V^2(r) S(qr) dr \ ,$$

(2.8)

$$S(qr) = 9 \frac{[\sin(qr) - qr \cos(qr)]^2}{(qr)^6} \ ,$$

(2.9)

where N is the number of particles and r their radius, $V(r) = 4\pi r^3/3$ their volume, n_A the number density of atom A within the particles, and $P(r)$ the normalized size distribution of the particles. f_M and n_M are the average atomic scattering factor and the number density of the surrounding matrix area.

There is, however, an additional scattering contribution from the pores in the carbon support, which has to be subtracted prior to any further data evaluation, i.e.,

$$I_M(q) = 2\pi(n_{\text{electrolyte}} f_{\text{electrolyte}} - n_M f_M)$$

(2.10)

$$\times (n_{\text{electrolyte}} f_{\text{electrolyte}} - n_M f_M)^* A_{\text{pore}} q^{-4} \ .$$

The q^{-4} dependence of the above contribution reflects Porod's law, which holds in the measured q range for pores with pore radii $r_{\text{pore}} > 4/q$. The scattering contrast results from the scattering factors $f_{\text{electrolyte}}$, f_M and number densities $n_{\text{electrolyte}}$, n_M of the atoms in the electrolyte and the matrix, respectively.

If the SAXS intensities are measured using X-rays with incident energies close to the Pt absorption edge, the atomic scattering amplitude f_{Pt} is varied by the anomalous scattering effect of Pt. This results in a decreased scattering intensity from the Pt particles, whereas the pore scattering terms remain unchanged, because the atomic scattering factors in the carbon and electrolyte matrix have their absorption edge at much lower X-ray energies. Therefore, the contribution of the Pt particles can be straightforwardly separated from the measured intensity contrast, using the common approximation of $n_{Pt}^2 f_{Pt}^2$ for $(n_{Pt}f_{Pt} - n_M f_M)(n_{Pt}f_{Pt} - n_M f_M)^*$:

$$I_{\text{particle}}(q, \varepsilon_1) - I_{\text{particle}}(q, \varepsilon_2) \tag{2.11}$$
$$= n_{Pt}^2 \left[f_{Pt}^2(\varepsilon_1) - f_{Pt}^2(\varepsilon_2) \right] \frac{N}{V_s} \int_0^\infty P(r) V^2(r) S(qr) \mathrm{d}r \ .$$

Using this equation, we can readily determine the distribution of the Pt particles and their volumes. Figure 2.2 shows the example of anomalous SAXS results of carbon-supported 10 wt% Pt catalyst measured at the Pt L_{III} absorption edge [16]. The intensity profile at the smaller q region, is the contribution of the pores in the carbon support, since no energy dependence is found. Scattering contrast, however, was obtained at the larger q region. This arises from the catalyst scattering, which represents the q dependence corresponding to the nanometer size of small Pt particles. The solid line in Fig. 2.2 is a fit of a log-normal size distribution of the mean radius $r_0 = 0.86$ nm and mean particle volume 3.7 nm^3. The present result clearly demonstrates that the anomalous SAXS technique is useful for characterizing individual sizes and distributions of nanometer-sized metal catalysts on porous supports by obtaining the scattering contrast due to the anomalous dispersion effect of metal particles.

The anomalous SAXS technique was also applied to the analysis of microcrystalline $CdS_x Se_{1-x}$ in silicate glasses. The optical properties of many oxide glasses are determined by particle microphases separated from the formerly homogeneous glass as a result of thermally induced diffusive decomposition in a supersaturated solution. In particular, silica glasses synthesized with around 1% $CdS_x Se_{1-x}$ have attracted recent interest because of their semiconducting behavior. At small crystalline sizes, quantum confinement effects can occur and the physical properties of the glasses with such microcrystalline phases differ considerably from those of corresponding bulk materials. Since the variation of interesting properties is known to depend strongly on size distribution, several techniques such as high-resolution transmission electron microscopy, small-angle X-ray scattering and Raman scattering have been employed. However, determination of the particle size by Raman scattering is a relativistic method based on the measurement of the frequency of confined acoustic phonons, and the results obtained are probably overestimated.

On the other hand, SAXS is quite useful for confirming the size distribution of the microcrystals qualitatively, and further introduction of

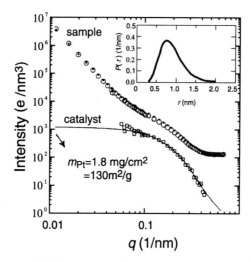

Fig. 2.2. Anomalous SXAS intensity of carbon-supported 10 wt% Pt catalyst. The scattering intensity contrast measured at the energies $\varepsilon_1 = 10\,353\,\text{eV}$ and $\varepsilon_2 = 11\,548\,\text{eV}$ near the Pt L_{III} absorption edge must be attributed to the scattering of Pt catalyst particles [16]. The *solid line* is a fit of a log-normal size distribution. The *insert* indicates the resultant log-normal size distribution of the Pt catalyst particles

anomalous SAXS can be used to examine the number density ratio of the crystalline phase and matrix n_A/n_M. Further consideration of equation (2.8) suggests that the square root of the observed intensity at $q = 0$, $\sqrt{I_{\text{particle}}(q = 0)}$ is approximated by a linear equation in the scattering contrast $\sqrt{(n_A f_A - n_M f_M)(n_A f_A - n_M f_M)^*}$. Therefore, the quantity n_A/n_M can be obtained straightforwardly by analyzing the measured intensity variation as a function of f_A. Goerigk et al. [17] applied this idea to the analysis of the $CdS_x Se_{1-x}$ nanocrystallite in silicate glasses and measured the variation of $\sqrt{I_{\text{particle}}(q = 0)}$ at the Se K absorption edge, as shown in Fig. 2.3. Least-squares analysis for the square root of the measured intensity using four different X-ray energies yielded the ratio $n_{CdS_x Se_{1-x}}/n_{SiO_2}$ together with the volume fraction of $CdS_x Se_{1-x}$. These structural parameters agree with the values expected from the chemical composition.

2.3.2 Nature of the Scattering Tail

A large, rapidly decreasing intensity called the scattering tail is commonly present in the SAXS patterns of metallic alloys. The origin of such tails is generally attributed either to internal particles or to surface heterogeneities. As an example, such a tail is suggested to be due to precipitation in the coarse grain boundary, in the case of phase separated alloys with medium-range SAXS signals. On the other hand, contamination and surface heterogeneities

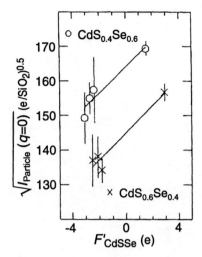

Fig. 2.3. Contrast variation for CdS_xSe_{1-x}-doped silicate glasses ($x = 0.4$ and 0.6) measured at the Se K absorption edge [17]. The scattering intensities were extrapolated to $q = 0$ and plotted as a function of

$$F'_{CdSSe} = 0.96f'_{Se} + 0.398f''_{Se} + \frac{x}{1-x}(0.96f'_S + 0.398f'_S) + \frac{1}{1-x}(0.96f'_{Cd} + 0.398f'_{Cd})$$

are the most feasible scatterers where there is no homogeneous precipitation, e.g., in as-quenched alloys or metallic glasses. The contrast between the material and its surroundings is, however, much stronger than that between the matrix and particles, so that very small numbers of superficial defects may give equivalent or sometimes stronger signals. By making available the scattering intensity contrast due to the anomalous scattering factor of a labeled atom, anomalous SAXS is quite useful for determining the origin of such scattering tails.

The pioneering anomalous SAXS study with respect to the scattering tail concerned the Cu–Ni–Sn alloy system [18, 19]. Figure 2.4a shows SAXS intensities of supersaturated Cu-15 wt%–Ni-8 wt%–Sn alloy annealed at 623 K for different times. Figure 2.4b gives the variation in scattering intensities of the sample annealed for 600 s in the vicinity of the Cu K absorption edge. As shown in Fig. 2.4b, the scattering intensities decrease when the incident energies approach the Cu K absorption edge from 8041 to 8972 eV. Such anomalous SAXS results, coupled with the observation of Ni- and Sn-rich precipitates at grain boundaries in as-quenched Cu-15 wt%–Ni-8 wt%–Sn alloy sample [20], allow us to conclude that the scattering tail close to the origin can be attributed to precipitates with considerably lower electronic density than that of the Cu-rich matrix. In these contexts, pure Sn is the inferred precipitate and the assumption of a phase such as $(Cu_xNi_{1-x})_3Sn$ or Ni_3Sn is readily excluded [19]. It is worth noting that the exact composition of the

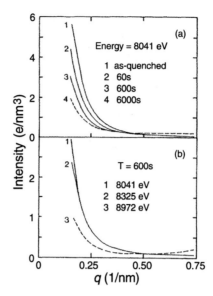

Fig. 2.4. Scattering for Cu-15 wt%–Ni-8 wt%–Sn alloy very close to the origin as a function of (**a**) aging time at 623 K and (**b**) incident energy [18]

precipitates cannot be determined from the presently available data alone, despite the fact that the assumption of Sn precipitates at grain boundaries is qualitatively consistent with the anomalous SAXS results of Fig. 2.4b. Nevertheless, the present anomalous SAXS results readily confirm the high Sn content and low electronic density of the precipitates.

Simon and his colleagues [21, 22] investigated alloy systems of Cu–Ni–Fe and Fe–Ni using the anomalous SAXS method. They specified the possible origins for the scattering tail near the origin. Anomalous SAXS patterns of $Cu_{0.425}Ni_{0.425}Fe_{0.15}$ alloy measured with energies near the Fe and Ni absorption edge are shown in Fig. 2.5. The scattering tails are hardly distinguishable from one energy to another near the angular origin. This suggests that the SAXS tail cannot be attributed to large-scale phase separation between Cu and NiFe, which usually gives intensity variations by a factor of 2.5 at both absorption edge. According to Simon and Lyon [21], the anomalous SAXS results on the tail of $Cu_{0.425}Ni_{0.425}Fe_{0.15}$ alloy are compatible with surface defects. In the simplest hypothesis of superficial roughness, the SXAS contrast is proportional to the square of the mean scattering factor of the sample f^2_{sample} and its value decreases only slightly as the incident energies approach the Fe and Ni edges.

As shown in Fig. 2.6, the scattering tail of $Fe_{0.56}Ni_{0.44}$ alloy measured at the Fe K absorption edge is not intense, being an order of magnitude weaker than the corresponding intensity of Cu–Ni–Fe. Nevertheless, the shape of this tail obeys the asymptotic Porod law, suggesting 'defects' with sharp

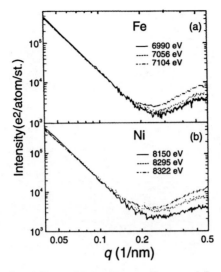

Fig. 2.5. Intensity of the $Cu_{0.425}Ni_{0.425}Fe_{0.15}$ alloy, aged for 201.6 ks at 773 K, as a function of the scattering wave vector for different energies: **(a)** near the Fe K absorption edge (7 112 eV), **(b)** near the Ni K absorption edge (8 333 eV) [21]

interfaces. The SAXS intensities decrease as incident X-ray energies are tuned closer to the absorption edge from 6 875 to 7 090 eV. Analysis of the scattering contrast $n_M f_M - n_d f_d$, where subscripts M and d indicate the matrix and the defect, respectively, readily excludes defect compounds such as Fe oxide and Fe carbide. Therefore, residual insoluble particles, which cause no anomalous scattering effect at the Fe K absorption edge, are candidate compounds for the origin of the scattering tail.

2.3.3 Decomposition Process

Numerous studies considering the decomposition kinetics of the Cu–Co alloy system have been carried out, because the small lattice misfit between Cu and Co allows us to consider this alloy as one of the ideal systems for the study of solid–solid nucleation. Decomposition of alloys with Co concentrations up to about 2.5 wt% has been discussed on the basis of the classical nucleation theory [23, 24]. However, recent reports show a discrepancy between the smallest measured particle radii and those of critical nuclei predicted by nucleation theory [25]. With a two-phase model, the Cu–Co system separates into a randomly distributed Co precipitate with sharp interfaces surrounded by the matrix. The latter consists mainly of Cu with a small amount of homogeneously distributed Co. When c_{Co}^P and c_{Co}^M are the Co concentrations in the precipitated phase (P) and the matrix (M), the SAXS intensity may be found from (2.8) with the contrast between $c_{Co}^P f_{Co} + (1 - c_{Co}^P) f_{Cu}$ and $c_{Co}^M f_{Co} + (1 - c_{Co}^M) f_{Cu}$.

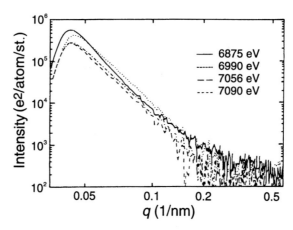

Fig. 2.6. Intensity of the $Fe_{0.56}Ni_{0.44}$ Invar alloy, aged for 421.2 ks at 873 K as a function of scattering wave vector for different energies at the Fe K absorption edge [21]

The study of Co precipitates is usually difficult because the difference between the atomic scattering factors of Cu and Co ($Z_{Co} = 27$ and $Z_{Cu} = 29$) is low in the case of ordinary X-ray scattering. This prompted the development of anomalous SAXS, in order to exploit enough contrast between the Co precipitate and the matrix to analyze the size distribution and volume fraction of Co precipitates [26,27]. Figure 2.7 shows the results of the anomalous SAXS measurement of the Cu–0.5 wt%Co sample annealed at 813 K for 27 min [26]. The difference in the scattering curves in Fig. 2.7a is caused by Co precipitates and the intensity contrast of the two scattering curves yields the separated scattering curve of Co precipitates alone. Further application of the theory [28,29] leads us to discuss the radius distribution and volume fraction of Co precipitates together with a variety of physical constants such as interfacial energy and diffusion coefficient. It should be noted that the extended compositional fluctuations model preceding the formation of stable Co precipitates [30] was excluded by the present anomalous SAXS measurement in the wave vector range between $q = 0.4$ and $5\,nm^{-1}$.

Supersaturated solid solutions of Cu–Ni–Sn alloy with a high content of Ni and Sn can be substantially hardened by a decomposition process during an aging treatment. This hardening mechanism of Cu–Ni–Sn alloys was classically attributed to the precipitation of the $(Cu, Ni)_3Sn$ phase or σ-Cu_4Sn [31–33]. However, several researchers have recently suggested that the spinodal decomposition of the α-phase in the temperature range 573–673 K is responsible for the age hardening [34–36]. In order to elucidate this age hardening behavior, X-ray diffraction and transmission electron microscopy (TEM) were used to follow the side-band evolution of the diffraction peak during the decomposition process and to detect the precipitation occurring

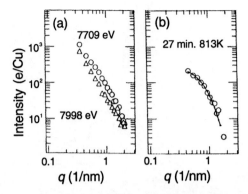

Fig. 2.7. (a) The SAXS at energies 7 709 and 7 998 eV. (b) The difference in the scattering curves caused by the scattering of Co precipitates was separated and fitted by a model function [27]

at the different decomposition stages. Wide-angle X-ray diffraction, in particular, is a sensitive technique for studying the beginning of the decomposition process. However, it should be noted that wide-angle X-ray diffraction is very sensitive to strain effects and is commonly enhanced by a sharp fundamental diffraction peak. On the other hand, the SAXS method is a less ambiguous technique for studying the early stages of a decomposition process because the SAXS pattern is almost unaffected by lattice displacements or strain effects. The spectrum of the composition waves and the square of their amplitudes, which are the structural parameters differing from those obtained by wide-angle X-ray diffraction, can be analyzed from the SAXS data.

Goudeau, Naudon and Welter [19] studied the decomposition process of Cu-15 wt%–Ni-8 wt%–Sn using the anomalous SAXS method in order to clear up this controversial problem of whether age hardening is caused by the nucleation and growth mechanism or by the spinodal decomposition process. Figure 2.8 shows the anomalous SAXS patterns measured for samples prepared by three different aging times: 60, 600 and 6 000 s, using four different energies near the absorption edges of Cu and Ni. The anomalous SAXS intensities due to in-grain decomposition decrease when the X-ray energy approaches the energy of the Ni K absorption edge from the lower energy side. On the contrary, a very sharp increase in the maximum intensity is observed at the Cu K absorption edge. The anomalous dispersion effect arising from Cu or Ni is clear, and the observed intensity variation includes the selective contribution of each element.

Generally, the decomposition of supersaturated solid solutions can be divided into two regimes:

– nucleation and growth, for which the instability results in a local perturbation, large in amplitude but small in extent,

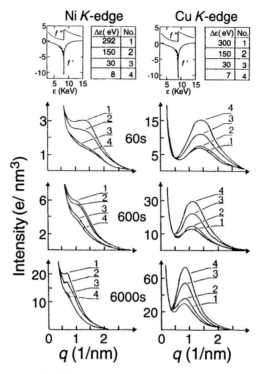

Fig. 2.8. Anomalous SAXS curves of Cu-15 wt%–Ni-8 wt%–Sn alloy measured on both Cu and Ni K absorption edges, aged at 623 K for 60, 600, 6 000 s. $\Delta\varepsilon$ denotes the difference between the incident X-ray energies from the absorption edge and corresponds to the number of curves [19]

– spinodal decomposition, which is considered as a perturbation throughout the volume of the sample, small in amplitude but large in extent.

Although the distinction between the two regimes is not clear in metallurgical systems, these can nevertheless be distinguished by analyzing the integrated intensity Q_0 calculated from

$$Q_0 = \frac{1}{2\pi^2} \int_{q_{\min}}^{q_c} q^2 I(q) \mathrm{d}q , \qquad (2.12)$$

where q_c and $q_{\min}$ are the experimental cutoff value and the minimum value, respectively. At first, Goudeau, Naudon and Welter [19] tried to calculate the integrated X-ray scattering intensity for Cu–Ni–Sn alloys based on a spinodal decomposition model, where the observed intensity is only proportional to the variation of the effective electronic density. However, fluctuations in Sn composition with approximately constant Cu/Ni ratio could not reproduce either the scattering contrast detected at both Ni and Cu K absorption edges or the aging behavior of the SAXS curves obtained experimentally. On the other hand, calculations based on a nucleation and growth mechanism can well reproduce the observed intensity variations of the present anomalous SAXS, using the simplest two-phase model of well defined precipitates of γ-phase embedded in a matrix of α-phase. The integrated intensity Q_0 is described by

$$Q_0 = \varphi(1 - \varphi)(n_p f_p - n_M f_M)(n_p f_p - n_M f_M)^* , \qquad (2.13)$$

where φ is the volume fraction of precipitates. Systematic calculation of equation (2.13) using the three variables of nickel content in the precipitates, the volume fraction of these precipitates together with the amount of Sn at the grain boundaries, leads to the conclusion that small precipitates of metastable $(Cu_{0.52}Ni_{0.48})_3Sn$ phase with DO_{22} structure are responsible for the SAXS signal. These results again indicate the usefulness of the anomalous SAXS method. In particular, this method is one of the most powerful tools for differentiating between competing models which otherwise fit equally well with the SAXS data in complex systems.

2.3.4 Amorphous Metals

A complete structural investigation of materials in any system must face the issue of phase separation. Nevertheless, the common wide-angle X-ray diffraction and electron microscopy often fail when they are applied to amorphous materials. Even SAXS, which is sensitive to electron density fluctuations on a scale of tens to hundreds of nanometers, cannot by itself distinguish such interesting phenomena from the cracks or voids. Here, anomalous SAXS is again a promising tool for analyzing fine-scale compositional modulations from other density fluctuations [37–40].

Metal–germanium amorphous films can be prepared typically from pure Ge to about 70 at.% metal. Such films exhibit physically interesting characteristics at higher metal concentrations. For example, superconductivity appears at 13 at.% Mo in the Mo–Ge alloy system, whereas ferromagnetism appears at 40 at.% Fe in the Fe–Ge system. In order to determine whether these interesting properties occur via phase separation or homogeneous alloy formation, anomalous SAXS experiments have been conducted for the amorphous metal–germanium alloys [38, 39].

Figure 2.9 shows the SAXS profiles of amorphous $Fe_x Ge_{100-x}$ alloys $(5 < x < 30)$ by illustrating a peak maximum near $2.8\,nm^{-1}$ which corresponds to a real space distance of the order of $2\,nm$. The position of the SAXS peak shifts towards the smaller wave vector region with increasing metal concentration, accompanied by an increase in the peak intensity. This indicates a variation of the correlation length of the composition fluctuation from 2 $(x = 5)$ to $14\,nm$ $(x = 30)$. These results clearly imply that very fine-scale composition modulation is likely to exist in the low metal concentration region of the amorphous Fe–Ge alloys and that its size scale increases with

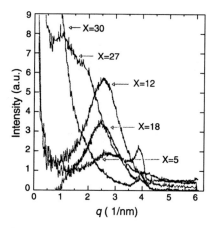

Fig. 2.9. Variations of peak intensity and position in amorphous Fe_xGe_{100-x} ($x = 5, 12, 18, 27$ and 30). The small narrow peak near $q = 4\,nm^{-1}$, which appears in several samples, is due to imperfect subtraction of the kapton substrate [38]

increasing metal content. As an example, Fig. 2.10 indicates the variation of the anomalous SAXS intensity of the amorphous $Fe_{12}Ge_{88}$ alloy, measured at energies below the K absorption edges of Fe and Ge. The SAXS intensity of the amorphous $Fe_{12}Ge_{88}$ alloy apparently decreases when the energies of the incident X-rays approach the Fe absorption edge from the lower energy side, whereas the SAXS intensity remains constant even though the energies of the incident X-rays increase toward the Ge absorption edge. The measured intensity variation allows us to conclude that Ge distributes rather uniformly and that the SAXS signal arises mainly from fluctuations with respect to the inhomogeneous distribution of Fe. Followed by the investigation of possible phases in the Fe–Ge system, the number density of Ge in pure Ge and in $FeGe_2$ is nearly the same. Therefore, phase separation of amorphous $M_{12}Ge_{88}$ alloy into amorphous Ge and an amorphous $FeGe_2$-like substance is quite feasible and consistent with a metal composition wave which permits the homogeneous Ge distribution.

Recently, two stages of glass transition temperatures as well as a wide supercooled liquid region were found in an amorphous $Zr_{33}Y_{27}Al_{15}Ni_{25}$ alloy [41]. The origin of this particular behavior has been attributed to the insoluble character of Y and Zr, and this suggests a structural inhomogeneity in the amorphous state. Figure 2.11 shows the anomalous SAXS profiles of the amorphous $Zr_{33}Y_{27}Al_{15}Ni_{25}$ alloy annealed at 773 K for 300 s, measured at Ni, Y and Zr K absorption edges by Sugiyama et al. [40]. The insert shows the SAXS data of an Al–Zn alloy as a reference sample, where no anomalous dispersion effect is detected. Although the positional resolution of anomalous SAXS signals is not sufficient, the anomalous dispersion effect is clearly observed, e.g., the remarkable decrease in the intensity for Ni and Zr

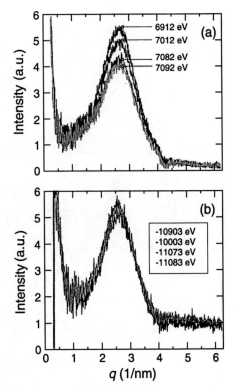

Fig. 2.10. Anomalous SAXS carves of amorphous $Fe_{12}Ge_{88}$ measured on (a) Fe and (b) Ge K absorption edges [38]

together with the increase for Y when the incident X-ray energy approaches the corresponding edges. These experimental results clearly exclude the possible precipitates Al_3Zr_5 as the origin of the SAXS peaks at about $0.65\,nm^{-1}$, although this crystalline phase is produced by annealing at $773\,K$ for more than $600\,s$.

Coupled with wide-angle anomalous X-ray scattering results suggesting the chemical fluctuation of Y-rich precipitates reported by Matsubara et al. [42], the amorphous Y–Al–Ni alloy or Y_3Al_2-like substance is likely to be a feasible scatterer for this annealed $Zr_{33}Y_{27}Al_{15}Ni_{25}$ alloy. A similar anomalous SAXS technique has been applied successfully to the study of amorphous CuTi and (Tb,Gd)Cu alloys [37, 43]. It is also interesting that $(Fe,Mn)_{35}Y_{65}$ metallic glass is a perfect representation of the segregation based on concentration fluctuations, and that the partial atomic volume ratio $\nu_{Fe,Mn}/\nu_Y$ could be obtained by anomalous SAXS [44].

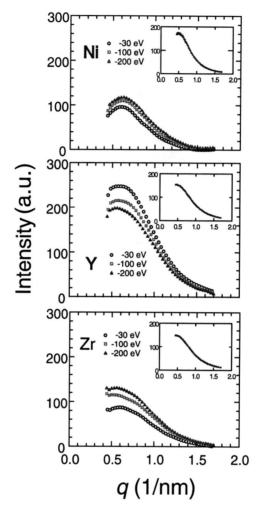

Fig. 2.11. Energy dependence of the SAXS intensity near the Ni, Y and Zr K absorption edges for the amorphous $Zr_{33}Y_{27}Al_{15}Ni_{25}$ alloy annealed at 773 K for 300 s [40]

2.3.5 Partial Structure Function

SAXS has been widely applied to the study of un-mixing in crystalline alloys. In the case of a ternary ABC alloy, the measured SAXS intensity profile is a linear combination of Fourier transforms of three independent pair correlation functions,

$$I(q) = \sum_{i,j=1}^{2} (f_i - f_M)(f_j - f_M)^* S_{ij}(q) , \qquad (2.14)$$

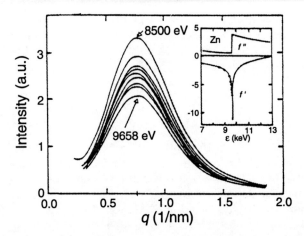

Fig. 2.12. Variation of SAXS intensity for the Al–9 at.%-Zn–9 at.%-Ag alloy, aged for 600 s at 398 K. The *upper* to *lower curves* were recorded at X-ray energies of 8 500, 9 000, 9 400, 9 560, 9 625, 9 640, 9 651 and 9 650 eV [48]. Since all energies are located below the absorption edge, the theoretical anomalous dispersion terms are given in the *insert*

where $S_{ij}(q)$ is the Fourier transform of the corresponding pair. Although (2.14) is simple, the extraction of the $S_{ij}(q)$ from experimental intensity data is not easy. The basic idea for solving (2.14) in terms of pair correlation functions is to perform three different experiments corresponding to three different values of the atomic scattering factors. In the case of neutron scattering, a variety of samples in the same metallurgical state, but prepared with different isotopic composition, are frequently used to obtain the $S_{ij}(q)$ from the measured intensity contrast. However, this method is limited by available numbers of isotopic elements, and its accuracy can be significantly affected by the sample preparation.

Anomalous SAXS is an alternative method, which can vary the scattering factor in the immediate neighborhood of the absorption edge. The feasibility of this anomalous SAXS method was well demonstrated by the study of the Guinier–Preston (GP) zones in phase-separated Al–Zn alloys, where strong energy dependence appears in the intensity profiles at the Zn absorption edge [45, 46]. This new approach has been extended to Al–Zn–Ag ternary alloys by Lyon and his colleagues [47, 48]. Variation of SAXS intensity for an $Al_{82}Zn_9Ag_9$ alloy aged at 398 K for 600 s is shown in Fig. 2.12. When the energy increases from 8 500 to 9 658 eV, the SAXS intensity appears to decrease monotonically due to the anomalous dispersion effect of Zn. Although only the weighted sums of the partial structure functions were obtained by the anomalous SAXS study for a single element, the results are consistent with the so-called two-phase un-mixing model from which the direction of tie lines of the miscibility gap can be determined. In this two-phase un-mixing model, sharp interface precipitates are embedded in a depleted matrix, while both phases have uniform concentrations, c_i^P and c_i^M, respectively. Then all partial structure factors have the same shape $S_{PM}(q)$ and (2.14) can be approximated using $\Delta c_i = c_i^P - c_i^M$ and $F_i = f_i - f_0$:

$$I(q) = \Big[\sum_{i,j}^{2} \Delta c_i \Delta c_j F_i F_j \Big]^2 S_{PM}(q) \ . \tag{2.15}$$

A simple way of analyzing the anomalous scattering data is to rewrite (2.15) as $[\Delta c_{Zn} F_{Zn} + \Delta c_{Ag} F_{Ag}]^2 S_{PM}(q)$. Then, the plot of $\sqrt{I(q)}$ against F_{Zn} should be linear, and the ratio of $\Delta c_{Ag}/\Delta c_{Zn}$ can be obtained from it, since the value of F_{Zn} changes as a function of the X-ray energy. The obtained composition ratio for $Al_{82}Zn_9Ag_9$, aged at $253\,K$ for $15\,days$, changes from 0.75 to 1.05 by a second aging at $398\,K$. This suggests that the clusters are initially richer in Zn than in Ag and that a more balanced composition relation was reached in the later stages of aging. Including the result that the ratio for those aged at $415\,K$ is fairly constant at about 1.1, the long-time aging of $Al_{82}Zn_9Ag_9$ alloy appears to be governed by the formation of one type of GP zone. On the other hand, this ratio drastically changes from 0.47 to 0.25 in the case of $Al_{88}Zn_{14}Ag_4$ alloy as the precipitates transform from GP zones to the stable ε' precipitate particles. These interesting conclusions on the slope of the tie-line can be obtained only when anomalous SAXS data is available. The technique has also been applied to the study of Fe–Ni–Co–Mo [49], Fe–Mn–Co–Mo [50], Ti–Al–Mo [51] and Ti–Nb–Al alloys [52].

Several ternary alloy systems such as Cu- and Ni-rich Cu–Ni–Fe alloys were considered as if they were pseudo-binaries. In this case, the anomalous SAXS method is one way to obtain partial structural information. It is then possible to check the validity of the pseudo-binary un-mixing hypothesis, which may provide useful information on the different solute partitioning.

Simon and Lyon [53–55] studied the partial structure factors of $Cu_{0.43}Ni_{0.42}Fe_{0.15}$ alloys annealed at $773\,K$, using the anomalous SAXS method. Figure 2.13 gives the anomalous SAXS intensities at the K absorption edges of Fe and Ni. The observed intensities are found to increase proportionally when approaching the edge. This shows the increase in Fe or Ni contrast with respect to Cu. The detected variation in the anomalous SAXS intensities is directly attributed to the partial structure functions of $S_{FeFe}(q)$, $S_{NiNi}(q)$, and $S_{FeNi}(q)$ for Fe–Fe, Fe–Ni and Ni–Ni pairs, respectively, together with the constraint equation $S_{CuCu}(q) = S_{FeFe}(q) + S_{NiNi}(q) + 2S_{FeNi}(q)$. The method for extracting the $S_{ij}(q)$ from measured intensity data is mathematically simple, but plagued by experimental uncertainties, such as calibration counting statistics and the relatively small variation in the atomic scattering factor owing to a less well conditioned matrix of linear equations, and this usually prevents us from obtaining reliable $S_{ij}(q)$ values. Nevertheless, this empirical difficulty can be circumvented by taking advantage of a number of linear equations obtained by recording several anomalous SAXS

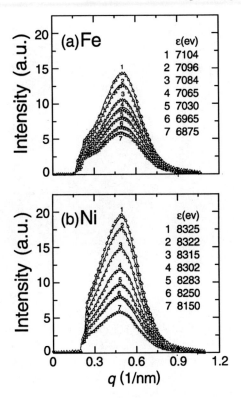

Fig. 2.13. Corrected experimental intensity of $Cu_{0.43}Ni_{0.42}Fe_{0.15}$ annealed at 773 K and best fit from the calculated partial structural functions: (**a**) near the Fe K absorption edge, (**b**) near the Ni K absorption edge [54]

measurements. Starting from reasonable initial $S_{ij}(q)$ values, several iteration procedures are applied on the basis of the gradient method, so as to minimize the residual function

$$N_q(S_{ij}) = \left| I(q, \varepsilon_k) - \sum_{i,j} F_i(\varepsilon_k) F_j^*(\varepsilon_k) S_{ij}(q) \right|^2. \tag{2.16}$$

The resultant partial structural functions are illustrated in Fig. 2.14. Maxima are found in the partial structure factors $S_{FeFe}(q)$, $S_{NiNi}(q)$ and $S_{CuCu}(q)$ from the like atom pairs and these patterns have a shape similar to the measured intensity curve. The ratios $S_{NiNi}(q)/S_{CuCu}(q)$ and $S_{FeFe}(q)/S_{CuCu}(q)$ are found to be constant, while the unlike-pair $S_{NiFe}(q)$ is close to zero over the whole q range. These interesting results suggest that there is a phase separation between a Cu-rich phase and a Ni–Fe phase, although there is long-range partitioning which kills the significant correlation between Ni and Fe in the latter domain. However, the value of the tie-line slope determined by anomalous SAXS deviates from one edge to another and the condition

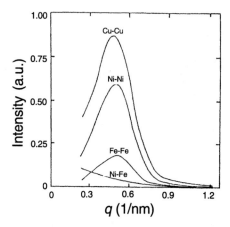

Fig. 2.14. Calculated partial structure functions of $Cu_{0.43}Ni_{0.42}Fe_{0.15}$ alloys annealed at 773 K using data in Fig. 2.13 [54]

$S_{ij}^2(q) - S_{ii}(q)S_{jj}(q) = 0$ for the pseudo-binary model as represented by equation (2.15) was not satisfied. These features clearly suggest a deviation from the pseudo-binary model. Further analysis was required for understanding experimental partial structure functions. Separating the partial structure functions into two contributions, a pseudo-binary contribution and a complementary one [55], confirmed Ni or Fe segregation at the interfaces between the Cu-rich phase and the Ni–Fe phase.

The partial structure functions estimated from anomalous SAXS measurements have also been reported in Fe–Ni–W [56] and Fe–Cr–Co alloys [57]. The latter case is well approximated by the pseudo-binary model.

2.3.6 Metallic Multilayers

New synthetic substances with multilayered structures such as metallic superlattice films have stimulated growing interest in recent years. They provide an ideal platform to study a variety of physical phenomena, such as superconductivity and magnetism. In particular, these new materials display great potential as optical elements for the soft X-ray and extreme ultraviolet region of the electromagnetic spectrum, as quasi-Bragg reflectors. For this application, interfaces play a crucial role in achieving optimum performance.

Several techniques have been developed for investigating the composition profiles of multilayers, although small-angle X-ray diffraction in the reflection geometry is classified as one of the most reliable tools for characterizing the periodicity and composition profiles of metallic multilayers [58]. The peak intensity for a multilayer sample consisting of two elements A and B layered alternately N times with period L is given by

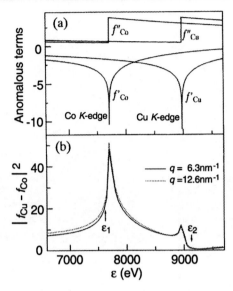

Fig. 2.15. (a) Theoretical anomalous dispersion terms of Co and Cu near their K absorption edges. (b) Energy dependence of $|f_{Cu} - f_{Co}|^2$ in the same energy region at $q = 6.3$ and $12.6 \, \text{nm}^{-1}$ [61]

$$I(q) = F(q)F^*(q)\frac{\sin^2(NqL/2)}{\sin^2(qL/2)} \ . \tag{2.17}$$

This intensity shows a sharp maximum at positions $q = 2\pi n/L$, where n is an integer and the corresponding structure factor of the n-th order peak is given by using the concentration profile $c(x)$ along the depth x:

$$F_n = 2(n_A f_A - n_B f_B)\int_0^{L/2} c(x)\cos\frac{2\pi n x}{L}\mathrm{d}x \ . \tag{2.18}$$

Since the intensities of diffraction peaks are proportional to the square of the difference between the average X-ray atomic scattering factors of the corresponding layers, structural characterization of some samples composed of layers with nearly equal atomic number elements is fundamentally difficult. However, this inconvenience can be effectively overcome using the anomalous scattering effect [59–61]. Figure 2.15 shows the variation of anomalous dispersion terms for Cu and Co at their K absorption edges, as an example. Due to the large variation of the anomalous dispersion terms, the peak intensity shows a distinct change near their absorption edges. Figure 2.16 shows the results of a difference between two intensity profiles of the Cu/Co multilayer measured at two different energies, 7 690 and 9 200 eV [61]. Although the higher order peaks are generally weak and occur against a very large background intensity, the present differential anomalous small-angle X-ray diffraction method can reduce such difficulties by precisely determining the

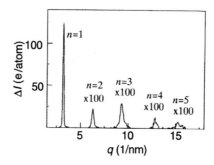

Fig. 2.16. Peak profiles of the Cu/Co multilayer obtained from intensities at 7 690 and 9 200 eV [61]

background intensity profile. This enables us to detect even the very weak peak at $n = 5$.

The concentration profile $c(x)$ is obtained by the Fourier transformation of the square root of the differential integrated intensity of Fig. 2.16 using

$$c(x) = \frac{2}{L} \sum_{n=1} \frac{|\Delta F_n| \phi_n}{|\Delta f|} \cos \frac{2\pi n x}{L} , \qquad (2.19)$$

where

$$|\Delta F_n| = \sqrt{|F_n(\varepsilon_1)|^2 - |F_n(\varepsilon_2)|^2} \qquad (2.20)$$

$$|\Delta f| = \sqrt{|n_A f_A(\varepsilon_1) - n_B f_B(\varepsilon_1)|^2 - |n_A f_A(\varepsilon_2) - n_B f_B(\varepsilon_2)|^2} . \quad (2.21)$$

Figure 2.17 shows the concentration profile $c(x)$ for the Cu/Co multilayer calculated from the differential AXS profile using (2.19), (2.20) and (2.21). By fitting the experimental $c(x)$ with trapezoidal profile, it is found that the periodic thickness is 2.045 ± 0.020 nm, while the thicknesses of the Cu and Co layers are 0.455 ± 0.005 nm and 0.500 ± 0.005 nm, respectively. It should be added that the phase factors ϕ_n were estimated so that the experimental integrated intensities were consistent with those obtained by inverse Fourier transformation of $c(x)$.

Such multilayered structures have also been applied as model systems to study interdiffusion on very short length and time scales [62]. In these studies, the following equation regarding the X-ray intensity diffracted at the modulation wave vector q is usually employed, in order to measure the diffusion coefficient in multilayered structures:

$$In \left[\frac{I(q)}{I_0} \right] = -2Dq^2 t , \qquad (2.22)$$

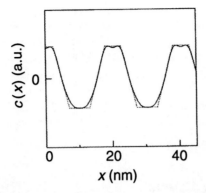

Fig. 2.17. Concentration profile in the Cu/Co multilayer obtained by Fourier transformation of the relative integrated intensities in Fig. 2.16 [61]

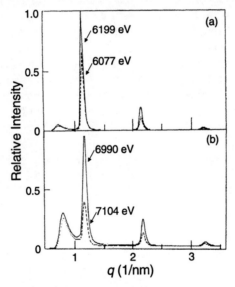

Fig. 2.18. Intensity profile of the Nd/Fe multilayer ($d_{Nd} = 3.7$ nm, $d_{Fe} = 1.6$ nm) measured (**a**) near the L_{III} absorption edge of Nd and (**b**) near the K absorption edge of Fe [63]

where I_0 and t denote the reference X-ray intensity and time, respectively. However, this simple analysis relies on the assumption that the two components of the multilayer sample have the same specific volume. For this reason, determination of the specific volumes in multilayers is essential. Simon et al. [63] proposed a new technique using anomalous SAXS. The integrated intensity may be almost proportional to the square of the difference of electron densities $|n_A f_A - n_B f_B|^2$. Therefore, the ratio of the specific volumes

n_A/n_B can be estimated from the intensity variation measured at energies close to the absorption edge of a constituent. Figure 2.18 shows the intensity profiles of an Nd–Fe multilayer measured at the L_{III} and K absorption edges of Nd and Fe, respectively. In both cases, the intensity variations corresponding to the change in anomalous dispersion terms are readily detected, and this anomalous SAXS method was enough to estimate the specific volume ratio with an experimental uncertainly of 5% [63]. This method has recently been extended to study atomic relaxations occurring during Nd–Fe interdiffusion [64].

2.4 Concluding Remarks

Small-angle X-ray scattering (SAXS) has been widely employed for characterizing inhomogeneities of materials in a variety of states. However, the required structural information about interesting particles is sometimes concealed by strong scattering from the substrate. It has also been found that several controversial models can fit equally well to the measured scattering data and this blocks further analytical investigation. Anomalous SAXS in the vicinity of the absorption edge of a constituent element in a sample enables one to obtain real structural images through the environmental information around a labeled element, as we have shown in this article through a selection of examples. Thus, the combination of SAXS and anomalous SAXS appears to provide useful answers to questions that remain unsolved using conventional SAXS data. It may also be noted that one can obtain even more information on systems including neighboring elements in the periodic table by clever use of anomalous X-ray scattering.

The recent development of synchrotron facilities with suitable experimental setups has opened the way to a wider use of SAXS in a variety of research fields. One example is X-ray magnetic circular dichroism (X-MCD). The dependence of the absorption of circularly polarized X-rays on the projection of the magnetization onto the photon propagation direction can provide the size distribution and correlation length of magnetic precipitates in the amorphous GdFe system [65]. Careful design of a SAXS spectrometer with an ultrahigh vacuum flight path, allows us to perform anomalous SAXS measurements at the K edges of relatively light elements [12,13]. Additionally, the development of a two-dimensional proportional detector (PSD) has prompted two-dimensional anomalous SAXS analysis of anisotropic nano-precipitates in a Cu–Ni–Fe single crystal [66]. Now, anomalous SAXS analysis is widely used as a result of the increased availability of intense, tunable X-rays with a high brilliance synchrotron radiation source. It would be interesting to extend the anomalous SAXS method to other subjects of structural characterization for materials, so that its usefulness and validity may be confirmed across a rather wider base.

References

1. A. Guinier and G. Fournet: *Small-Angle Scattering of X-rays* (John Wiley & Sons, New York, and Chapman & Hall, London 1955)
2. O. Glatter and O. Kratky: *Small-Angle X-Ray Scattering* (Academic press, London 1982)
3. H. Brumberger: *Modern Aspects of Small-Angle Scattering*, (Kluwer Academic, Dordrecht 1993)
4. R.W. James: *The Optical Principles of the Diffraction of X-rays* (G. Bell & Sons, London 1958)
5. Y. Waseda: *Novel Application of Anomalous (Resonance) X-ray Scattering for Structural Characterization of Disordered Materials* (Springer, Berlin, Heidelberg, New York 1984); SCM-AXS
   http://www.iamp.tohoku.ac.jp/database/scm/AXS
6. G. Materlik, C.J. Sparks, and K. Fischer: *Resonant Anomalous X-ray Scattering: Theory and Applications* (Elsevier Science, Oxford 1994)
7. D.T. Cromer and D. Liberman: J. Chem. Phys. **53**, 1891 (1970)
8. K. Shinoda, K. Sugiyama, and Y. Waseda: High Temp. Mat. Process. **14**, 75 (1995)
9. H.-G. Haubold, K. Gruenhagen, M. Wagener, H. Jungbluth, H. Heer, H. Pfeil, G. Rongen, G. Brandenberg, R. Moeller, J. Matzerath, P. Hiller, and H. Halling: Rev. Sci. Instrum. **60**, 1943 (1989)
10. G.G. Long, P.R. Jemian, J.R. Weertman, D.R. Black, H.E. Burdette, and R. Spal: J. Appl. Cryst. **24**, 30 (1991)
11. S. Wakatsuki, K.O. Hodgson, D. Eliezer, M. Rice, S. Hubbard, N. Gillis, S. Doniach, and U. Spann: Rev. Sci. Instrum. **63**, 1736 (1992)
12. H.B. Stuhrmann: Top. Curr. Chem. **145**, 151 (1988)
13. O. Kühnholz: J. Appl. Cryst. **24**, 811 (1991)
14. V. Luzzati: Acta Crystal. **13**, 939 (1960)
15. H.-G. Haubold and X.H. Wang: Nucl. Instrum. Methods. Phys. Rev. B **97**, 50 (1995)
16. H.-G. Haubold, X.H. Wang, G. Goerigk, and W. Schilling: J. Appl. Cryst. **30**, 653 (1997)
17. G. Goerigk, H.-G. Haubold, C. Klingshirn, and A. Uhrig: J. Appl. Cryst. **27**, 907 (1994)
18. P. Goudeau, A. Naudon, and J.-M. Welter: Scrip. Metal. **22**, 1019 (1998)
19. P. Goudeau, A. Naudon, and J.-M. Welter: J. Appl. Cryst. **23**, 266 (1990)
20. L.E. Collins and J.R. Barry: Mater. Sci. Eng. **98**, 335 (1998)
21. J.P. Simon and O. Lyon: J. Appl. Cryst. **24**, 1027 (1991)
22. J.P. Simon, O. Lyon, F. Faudot, L. Boulanger, and O. Dimitrov: Acta Metal. Mater. **40**, 2693 (1992)
23. I.S. Servi and D. Turnbull: Acta Metall. **14**, 161 (1966)
24. F.K. LeGoues and H.I. Aaronson: Acta Metall. **32**, 1855 (1984)
25. H. Wendt and P. Haasen: Scr. Metall. **19**, 1053 (1985)
26. G. Goerigk, H.-G. Haubold, and W. Schilling: J. Appl. Cryst. **30**, 1041 (1997)
27. P. Ancrenaz, C. Servant, and O. Lyon: Acta Crystallogr. B **49**, 458 (1993)
28. I.M. Lifshifz and V.V. Slyozov: J. Phys. Chem. Solids **19**, 35 (1961)
29. C. Wagner: Z. Elektrochem. **65**, 581 (1961)
30. W. Wagner: Acta Metall. Mater. **38**, 2711 (1990)

31. F.A. Badia: Proc. Am. Soc. Test. Mater. **62**, 665 (1962)
32. W. Léo: Metallurgy **21**, 908 (1967)
33. B.D. Bastow and D.H. Kirwood: J. Inst. Met. **99**, 277 (1971)
34. L.H. Schwartz, S. Mahajan, and J.T. Plewes: Acta Metall. **22**, 601 (1974)
35. B.G. Lefevre, A.T. D'annessa, and D. Kalish: Metall. Trans. **9** A, 577 (1978)
36. B. Ditchek and L.H. Schwartz: Acta Metall. **28**, 807 (1980)
37. P. Goudeau, A. Naudon, A. Chamberod, B. Rodmacq, and C.E. Williams: Europhys. Lett. **3**, 269 (1987)
38. M. Rice, S. Wakatsuki, and A. Bienenstock: J. Appl. Cryst. **24**, 598 (1991)
39. M.J. Regan and A. Bienenstock: Phys. Rev. B **51**, 12170 (1995)
40. K. Sugiyama, A.H. Shinohara, Y. Waseda, S. Chen, and A. Inoue: Mater Trans. JIM **35**, 481 (1994)
41. T. Zhang, A. Inoue, S. Chen, and T. Masumoto: Mater Trans. JIM **33**, 143 (1993)
42. E. Matsubara, K. Sugiyama, A.H. Shinohara, and Y. Waseda: Mat. Sci. Eng. A **179/180**, 444 (1994)
43. M. Maret, J.P. Simon, B. Boucher, R. Tourbot, and O. Lyon: J. Phys.: Condens. Matter **4**, 9709 (1992)
44. M. Maret, J.P. Simon, and O. Lyon: J. Phys.: Condens Matter **1**, 10249 (1989)
45. P. Goudeau, A. Naudon, A. Fontaine, and C.E. Williams: J. Physique Lett. **46**, L255 (1985)
46. P. Goudeau, A. Fontaine, A. Naudon, and C.E. Williams: J. Appl. Cryst. **19**, 19 (1986)
47. O. Lyon, J.J. Hoyt, R. Pro, B.E.C. Davis, B. Clark, D. de Fontaine, and J.P. Simon: J. Appl. Cryst. **18**, 480 (1985)
48. O. Lyon and J.P. Simon: Acta Metal. **34**, 1197 (1986)
49. N. Bouzid, C. Servant, and O. Lyon: Philos Mag. B **57**, 343 (1988)
50. C. Servant and N. Bouzid: Acta Metall. **36**, 2771 (1988)
51. S. Djanarthany, C. Servant, and O. Lyon: Philos. Mag. A **66**, 575 (1992)
52. F.A. Sadi and C. Servant: Philos. Mag. A **80**, 639 (2000)
53. J.P. Simon and O. Lyon: Philos. Mag. Lett. **55**, 75 (1987)
54. O. Lyon and J.P. Simon: Phys. Rev. B **35**, 5164 (1987)
55. O. Lyon and J.P. Simon: J. Phys. F.: Met. Phys. **18**, 1787 (1988)
56. C. Servant, O. Lyon, and J.P. Simon: Acta Metall. **37**, 2403 (1989)
57. J.P. Simon and O. Lyon: Acta Metall. **37**, 1727 (1989)
58. L.L. Chang and B.C. Giessen (Eds.): *Synthetic Modulated Structure* (Academic Press, New York 1985)
59. H.E. Fischer, H. Fischer, O. Durand, O. Pellegrino, S. Andrieu, M. Piecuch, S. Lefebvre, and M. Bessière: Nuclear Instrum. Meth. Phys. Res. B **97**, 402 (1995)
60. N. Nakayama, I. Moritani, T. Shinjo, Y. Fujii, and S. Sasaki: J. Phys. F: Met. Phys. **18**, 429 (1988)
61. K. Kato, E. Matsubara, M. Saito, T. Kosaka, Y. Waseda, and K. Inomata: Mater. Trans. JIM **36**, 408 (1995)
62. A. Bruson, M. Piecuch, and G. Marchal: J. Appl. Phys. **58**, 1229 (1985)
63. J.P. Simon, O. Lyon, A. Bruson, G. Marchal, and M.Piecuch: J. Appl. Cryst. **21**, 317 (1988)
64. J.P. Simon, O. Lyon, A. Bruson, F. Rieutord: J. Apply. Cryst. **24**, 156 (1991)
65. P. Fischer, R. Zeller, G. Schütz, G. Goerigk, H.-G. Haubold, K. Pruegl, and G. Bayreuther: J. Appl. Phys. **83**, 7088 (1998)
66. O. Lyon, I. Guillon, and C. Servant: J. Appl. Cryst. **34**, 484 (2001)

3 Icosahedral Clusters
in RE(TM$_{1-x}$Al$_x$)$_{13}$ Amorphous Alloys

K. Fukamichi, A. Fujita, T.H. Chiang, E. Matsubara, and Y. Waseda

Magnetic properties of $3d$ transition metals (TMs) and alloys are decided by the balance between the magnitude of the electron correlation and the kinetic energy of $3d$ electrons [1–4]. In other words, the magnetic properties are influenced by the atomic structures, because the transfer probability of $3d$ electrons is easily influenced by the interatomic distance and the coordination number [1,3], accompanied by the changes in magnitude of both the electron correlation and the kinetic energy of $3d$ electrons. Such features of $3d$ electrons are closely correlated to the origins of various important and attractive problems in magnetism. A unified model of itinerant- and localized-electron models by taking spin fluctuations (SFs) into consideration successfully explains various magnetic properties in crystalline homogeneous systems [2].

The relationship between structural disorder and electron correlations is one of many basic problems that need to be solved. The magnetic properties of $3d$ TM-based amorphous alloys are also influenced by itinerant features of $3d$ electrons. However, magnetic inhomogeneity due to structural and concentration fluctuation in amorphous alloys prevents one from precisely analyzing their magnetic properties, because SFs and concentration fluctuations oppose each other. Recently, however, amorphous structures with well-defined structural correlations have been confirmed in several La(TM$_{1-x}$Al$_x$)$_{13}$ amorphous alloy systems. In the present chapter, we discuss magnetic properties of these amorphous alloys in terms of SF theories.

This chapter is organized as follows. Section 3.1 will outline the relationship between the structure and magnetic properties of amorphous alloys. In Sect. 3.2, we present the characteristic amorphous structure of La(TM$_{1-x}$Al$_x$)$_{13}$ systems, together with that of La(Co$_{1-x}$Mn$_x$)$_{13}$. Theoretical aspects of magnetism in itinerant-electron ferromagnets are briefly reviewed in Sect. 3.3. Section 3.4 deals with the relationship between the fundamental magnetic properties and the characteristic atomic structure of RE(TM$_{1-x}$Al$_x$)$_{13}$ amorphous alloys by taking spin fluctuations (SFs) into consideration. In Sect. 3.5, the discussion will be mainly devoted to the remarkable magnetovolume effects associated with SFs in La(TM$_{1-x}$Al$_x$)$_{13}$ amorphous alloys. The magnetoelastic properties and linear saturation magnetostrictions are discussed in Sect. 3.6. In Sect. 3.7, we describe the significant effects of pressure on the magnetic properties of La(Fe$_{1-x}$Al$_x$)$_{13}$ amorphous alloys, these being closely connected with the remarkable magnetovolume and magnetoelastic effects. Section 3.8 contains a brief conclusion. For the sake of convenience, abbreviations and symbols are listed at the end of the chapter.

3.1 Atomic Structures and Magnetic Properties of Amorphous Alloys

The linear muffin-tin orbital (LMTO) method is considered to be the most useful calculation for the band structure of complicated crystals with many atoms in a unit cell. The electronic structures of amorphous materials can be calculated using the tight-binding LMTO method combined with the recursion method, if the amorphous structure model is given [5]. Recently, the electronic structures of amorphous alloys have been calculated theoretically on the basis of a geometrical-mean model for amorphous structures and transfer integrals [6]. A finite-temperature theory based on the functional integral method and the distribution function method have been proposed and these theories have been discussed by Kakehashi et al. [7]. In these theories, the d electron number N_d, the mean value of the density of states (DOS) $[\rho(\varepsilon)]_S$, the atomic-size difference, and the function of the structural fluctuation of the interatomic distance defined by the following expression are the key factors:

$$\Delta^{1/2} = \frac{\left[(\delta r)^2\right]_S^{1/2}}{[r]_S} , \tag{3.1}$$

where $[r]_s$ and $[(\delta r)^2]_s^{1/2}$ denote the average interatomic distance which is obtained from the first peak of the pair distribution function and its fluctuation which is obtained from the half width of the peak, respectively.

In amorphous alloys, $3d$ electrons exhibit itinerant characters and magnetic properties are strongly influenced by the atomic structures [5, 8–10]. Although the amorphous structure has no translational symmetry, the first nearest-neighbor atomic distance and the coordination number CN are distributed around each average value. Since $3d$ electrons are bound by the atomic potential, the short-range local environment of the atomic configuration dominates the motion of $3d$ electrons. Consequently, various magnetic properties in the amorphous state can be correlated to the interatomic distance and the coordination number of atoms [10]. It should be pointed out that the local environment of the amorphous structure is similar to that of fcc metals [5]. Furthermore, amorphous alloys contain large free volumes. The main peak of the DOS in $3d$ electron bands in the amorphous state is located at a lower energy than that of fcc metals, close to the Fermi level [10, 11]. Strong ferromagnetic properties in many Co-based amorphous alloys can be explained from this viewpoint, although detailed mechanisms such as orbital hybridization depend on the constituent elements and their concentrations [10].

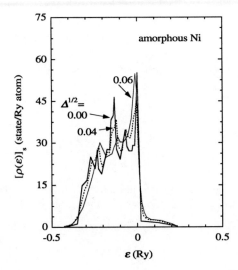

Fig. 3.1. Band structures of Ni in the amorphous state as a function of the structural fluctuation $\Delta^{1/2}$ [7]

The structural fluctuation defined by (3.1) has an obvious influence on magnetic properties in the regime where the magnitude of electron correlation and that of the kinetic energy of electrons are comparable. Spin-glass (SG) behavior appears in several Fe-based amorphous alloys [12, 13]. In Fe-based amorphous alloys, the main peak of the DOS is located on the higher energy side of the Fermi level ε_F [7] and the ferromagnetic phase is unstable in contrast to Co-based amorphous alloys. Accordingly, exchange interactions and the local moment become highly sensitive to the local environment in Fe-based amorphous alloys, i.e., the amplitude of the local moment is governed by the local environment and the sign of the exchange interaction changes with regard to the amplitude of the local moment [11], resulting in the spin-glass state. As a result, the spin-glass state in Fe-based amorphous alloys would be drastically changed by a slight change in the structural parameters.

Weak itinerant-electron (WIE) ferromagnetic properties have been observed in Ni-based amorphous alloys [14–16]. Taking the $3d$ electron number into consideration, the main peak of the DOS is expected to be located on the lower energy side of the Fermi level [11]. Comparing the band calculations, we should notice that the onset of ferromagnetism in pure Ni in the amorphous state is very sensitive to the interatomic distance r , the coordination number CN and the width of distribution δr, that is, the onset of ferromagnetism rests upon a slight change in the structural parameters [8–10].

Kakehashi et al. have demonstrated the change in the $3d$ bands of pure Ni in the amorphous state as a function of the structural fluctuation $\Delta^{1/2}$, as shown in Fig. 3.1 [7]. The resultant magnetic phase diagrams for the magnetic moment M_{Ni} and the Curie temperature T_C are shown in Figs. 3.2 and 3.3,

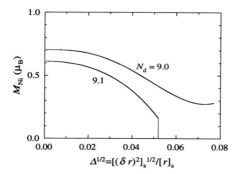

Fig. 3.2. Dependence of the magnetic moment M_{Ni} on the structural fluctuation $\Delta^{1/2}$ for different values of the number of $3d$ electron N_d [7]

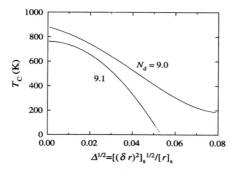

Fig. 3.3. Dependence of the Curie temperature T_C on the structural fluctuation $\Delta^{1/2}$ for different values of the number of $3d$ electron N_d [7]

where the numbers of $3d$ electrons N_d are assumed to be 9.0 and 9.1. The ratio $\Delta^{1/2}$ between the half maximum width of the distribution function δr and the average distance of the first nearest-neighbor Ni atoms is a measure of the degree of structural fluctuation. As seen from the figures, the ferromagnetic properties with $N_d = 9.0$ and 9.1 gradually weaken with increasing $\Delta^{1/2}$, although the decrement in M_{Ni} and T_C for each N_d differs in detail. This behavior is explained by the shift of the main peak of $3d$ bands to a lower energy region with increasing $\Delta^{1/2}$ [7], resulting in a decrease in the DOS of $3d$ bands at the Fermi level. Experimentally, δr and r interrelate with each other, and hence their relation is observed as a structural short-range order, especially for amorphous alloys containing $5d$ or $4f$ elements, because the atomic radius of these elements is much larger than that of $3d$ elements and the coordination with these elements significantly influences the interatomic distance [17]. Consequently, atomic structural analyses are very important when discussing the details of magnetic properties in amorphous alloys.

3.2 Structural Characteristics of La(TM$_{1-x}$Al$_x$)$_{13}$ Amorphous Alloys

The structures of La(TM$_{1-x}$Al$_x$)$_{13}$ amorphous alloys differ from those of binary metallic amorphous alloy systems. Before the discussion, we therefore briefly explain the crystal structure of La(TM$_{1-x}$Al$_x$)$_{13}$ compounds in order to obtain a deeper understanding of the amorphous structures.

3.2.1 NaZn$_{13}$-Type Crystal Structure

In the crystalline state, the La(Fe$_{1-x}$Al$_x$)$_{13}$ system has been investigated intensively. La(Fe$_{1-x}$Al$_x$)$_{13}$ compounds have a NaZn$_{13}$-type structure with $Fm\bar{3}c(O_h^6)$ space-group symmetry. Figure 3.4a shows the unit cell of the NaZn$_{13}$-type structure [18]. The conventional cubic cell is complicated and consists of 112 atoms, i.e., La on 8a, TM on 8b and TM(Al) on 96i sites, since the conventional cubic cell consists of eight small cubes, and there are two La(TM$_{1-x}$Al$_x$)$_{13}$ formula units in a primitive unit cell, i.e., 28 atoms with one kind of 2La atoms and two kinds of TM atoms, i.e., 2TM^I and 24TM^{II} atoms. The TM^I atom is located at the center of the icosahedron and the TM^{II} atom is located at the vertices of the icosahedron and composed of TM

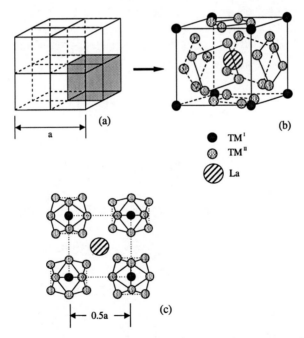

TM^I

TM^{II}

La

Fig. 3.4. Crystal structure of NaZn$_{13}$-type La(Fe$_{1-x}$Al$_x$)$_{13}$ compounds [18,19]. (**a**) conventional cubic cell, (**b**) primitive unit cell, (**c**) projection along the c axis of four TM^{II} icosahedra with the central TM^I atom

and Al atoms. The TM^{II} atoms are surrounded by 9 nearest TM^{II} atoms and 1 TM^{I} atom. The local environment of a TM^{I} is very similar to that of an fcc type, because it is surrounded by 12 TM^{II} atoms, as shown in Fig. 3.4c [19]. The La and TM^{I} atoms form a cubic CsCl(B2) structure, in which 4 TM^{II} atoms are located on each face of a cube, as shown in Fig. 3.4b [18]. The La atoms occupy the $(1/4, 1/4, 1/4)$ sites plus their image under symmetry operations, as illustrated in Fig. 3.4c.

Atomic structure parameters such as r and CN are given for the $La(Fe_{0.90}Al_{0.10})_{13}$ compound in Table 3.1. Fe^{I} and Fe^{II} correspond to TM in Fig. 3.4b. The intra-icosahedral cluster atomic correlations are underlined whilst the inter-icosahedral cluster atomic correlations are not. It should be noted that there are four kinds of nearest-neighbor correlation for Fe–Fe ranging from 0.247 to 0.286 nm.

3.2.2 $La(Fe_{1-x}Al_x)_{13}$ Amorphous Alloys

No stable crystalline Fe–La intermetallic compounds exist because the heat of alloying between Fe and La is positive. However, cubic $NaZn_{13}$-type $La(Fe_{1-x}Al_x)_{13}$ compounds have been successfully prepared by substituting Al for some Fe in RE–TM_{13} (RE rare-earth element, TM transition metal) [21]. The $La(Fe_{1-x}Al_x)_{13}$ compounds can be stabilized in the concentration range $0.08 \leq x \leq 0.54$, and the lattice constant of these intermetallic compounds decreases linearly with increasing Fe content [19]. The cubic crystal structure of $La(Fe_{1-x}Al_x)_{13}$ is composed of many icosahedral clusters.

Table 3.1. Atomic structure parameters of the $La(Fe_{0.90}Al_{0.10})_{13}$ compound. *Underlined terms* indicate intra-icosahedral cluster atomic correlations [20]

Pair	r [nm]	Correlation and coordination number
Fe–Fe	0.247–0.286	Fe^{I}–Fe^{II}(12), Fe^{II}–Fe^{I}(1), Fe^{II}–Fe^{II}(4), Fe^{II}–Fe^{II}(5)
Fe–Fe	0.352	Fe^{II}–Fe^{II}(1)
Fe–Fe	0.396–0.423	Fe^{I}–Fe^{II}(12), Fe^{II}–Fe^{I}(1), Fe^{II}–Fe^{II}(5), Fe^{II}–Fe^{II}(8)
Fe–Fe	0.462–0.465	Fe^{II}–Fe^{II}(8)
Fe–Fe	0.491–0.511	Fe^{I}–Fe^{II}(12), Fe^{II}–Fe^{I}(1), Fe^{II}–Fe^{II}(1), Fe^{II}–Fe^{II}(6)
Fe–Fe	0.545	Fe^{II}–Fe^{II}(4)
Fe–Fe	0.579–0.588	Fe^{I}–Fe^{I}(6), Fe^{I}–Fe^{II}(12), Fe^{II}–Fe^{I}(1), Fe^{II}–Fe^{II}(6)
La–Fe	0.339	La–Fe^{II}(24)
La–Fe	0.510–0.520	La–Fe^{I}(8), La–Fe^{II}(24)
La–La	0.579	La–La(8)

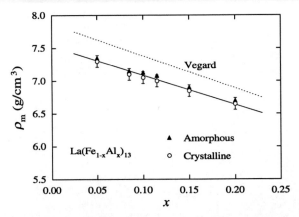

Fig. 3.5. Concentration dependence of the room-temperature mass density ρ_m for La(Fe$_{1-x}$Al$_x$)$_{13}$ amorphous alloys and their crystalline counterparts [35,36]

The room-temperature mass density ρ_m is sensitive to the characteristics of the atomic structures. Figure 3.5 shows the concentration dependence of ρ_m for the La(Fe$_{1-x}$Al$_x$)$_{13}$ amorphous alloys, together with that for the crystalline compounds. The broken line stands for Vegard's law [22]. In both the amorphous and crystalline states, ρ_m exhibits a gradual linear decrease with increasing x, showing lower values than those obtained from Vegard's law. The deviation from this law increases with decreasing x. Generally, the difference between ρ_m in the amorphous and crystalline states is less than 2% and the latter state is denser [23]. However, the point to note is that the La(Fe$_{1-x}$Al$_x$)$_{13}$ alloys in both states exhibit almost the same values, as shown in Fig. 3.5.

The reduced interference functions (RIFs) $Qi(Q)$ of the La(Fe$_{1-x}$Al$_x$)$_{13}$ amorphous alloys ($0.05 \leq x \leq 0.20$) are shown in Fig. 3.6 [24]. The observed intensities were corrected for air scattering, absorption and polarization and converted into absolute units per atom $I_{eu}(Q)$ by the generalized Krogh–Moe–Norman method [25] with the X-ray atomic scattering factors including the anomalous dispersion terms given in [186]. Then the coherent intensity $I_{eu}(Q)$ in absolute units was obtained by subtracting the theoretical Compton scattering given in [187] with the Breit–Dirac recoil factors. From the intensity $I_{eu}(Q)$, the RIF $Qi(Q)$ of the samples was evaluated using [17]

$$Qi(Q) = Q \frac{I_{eu}(Q) - \sum\limits_{i=1}^{N} c_i f_i^2}{(\sum\limits_{i=1}^{N} c_i f_i)^2} , \qquad (3.2)$$

where $Q = 4\pi \sin\theta/\lambda$, θ is half of the angle between the incident and diffracted beams, λ is the wavelength, and N is the total number of constituent elements. c_j and f_j are the atomic fraction and the atomic scat-

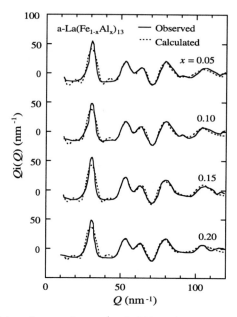

Fig. 3.6. Reduced interference functions $Qi(Q)$ of $La(Fe_{1-x}Al_x)_{13}$ amorphous alloys [24, 33]. *Solid line*: observation, *broken line*: calculated values

tering factor of the jth element, respectively. One of the notable features of the structural profiles in Fig. 3.6 is that fairly distinct oscillations occur even in the high-Q region. This kind of intensity profile is scarcely observed in metallic amorphous alloys and rather resembles that for oxide glasses, implying the presence of chemical short-range ordering (SRO) clusters with definite bond lengths and angular relations [26]. On the basis of the successful results obtained in determining the local unit structure of SiO_2 and BeF_2 glasses [27], the interatomic distances and the coordination numbers of these amorphous alloys were evaluated using the least-squares variational method for the interference function.

The nonlinear least-squares program [28] was slightly modified for the analysis. A brief description of this method required for the following discussion is given below. The interference function $Qi(Q)$ may be given by [28]

$$Qi(Q) = \sum_{j=1}^{N'} \sum_{i=1}^{N} N_{ij} \exp(-b_{ij}Q^2) \frac{f_i(Q)f_j(Q)}{\sum_{k=1}^{N} f_k(Q)} \frac{\sin(Qr_{ij})}{r_{ij}} \tag{3.3}$$

$$+ \sum_{\alpha=1}^{N'} \sum_{\beta=1}^{N} 4\pi\rho_0 c_\alpha c_\beta \exp(-b_{\alpha\beta}Q^2) \frac{f_\alpha(Q)f_\beta(Q)}{\sum_{k=1}^{N} f_k(Q)} \frac{Qr'_{\alpha\beta}\cos(Qr'_{\alpha\beta}) - \sin(Qr'_{\alpha\beta})}{Q^2} ,$$

where N' is the number of near-neighbor atoms for a type-j atom, N_{ij} and r_{ij} are the coordination number and the average distance of the i–j pairs, respectively, and b_{ij} is the mean-square variation. In this relation, the average number N_{ij} of the nearest j atoms around any origin atom of type i is assumed to be separated by an average distance r_{ij} and the distributions with respect to each correlation are expressed as discrete Gaussian-like functions with mean-square width $2b_{ij}$. The distributions of higher neighbor correlations are approximated by a continuous distribution with average number density ρ_0. The quantities $r'_{\alpha\beta}$ and $b'_{\alpha\beta}$ represent the average size and the variation of the boundary region, respectively. The parameters distinguished by α and β have the same meanings but for α and β atoms in the boundary region of the continuous distributions.

In practice, the distances and coordination numbers of near-neighbor correlations are obtained by the least-squares calculation of (3.3) so as to reproduce the experimental interference function data. The interference functions obtained in this procedure are shown by the dotted curves in Fig. 3.6. Generally speaking, icosahedral clusters tend to appear in the liquid metal and amorphous alloys [17, 29–31]. The icosahedral cluster is an isotropic and dense cluster, and atomic potential energy is gained by the formation of such clusters in amorphous metallic alloys [31]. Furthermore, local environments of amorphous alloys resemble those in the crystalline counterparts. Accordingly, the starting parameters for the least-squares variations were set by calculating the average coordination numbers and atomic distances in the La(Fe$_{0.91}$Al$_{0.09}$)$_{13}$ crystalline compound at 300 K [32] (see Table 3.1). These values are given in Table 3.2 with the final results for the La(Fe$_{0.90}$Al$_{0.10}$)$_{13}$ amorphous alloy. Errors in the results are estimated from the variance–covariance matrix in the least-squares variational method.

For this least-squares variational analysis, it is assumed that Fe and Al atoms randomly share atomic sites at the vertices of the icosahedral clusters in the amorphous structure as in the crystal structure. Since Fe and Al atoms randomly share the same atomic sites in the crystal structure [19], the coordination numbers for Fe–Al pairs are readily calculated simply by multiplying the ratio of the atomic fractions of Al and Fe, $x/(1-x)$, by the coordination number for Fe–Fe pairs in Table 3.2. In the La(Fe$_{0.91}$Al$_{0.09}$)$_{13}$ crystal, the icosahedral clusters formed by Fe and Al atoms located at their vertices and Fe atoms located at the center are present at the corner of the cubic cell, and the La atom is located at its body-centered position (see Fig. 3.4c). Since the CN and r values of Fe–Fe pairs in the amorphous alloy and in the crystal in Table 3.2 indicate little difference, it is plausible that local SRO clusters in the amorphous alloys are the same as those in the crystalline state [19, 32].

The formation of icosahedral clusters gives noble characteristics in the radial distribution function (RDF). The RDF is calculated by the Fourier transformation of the interference functions,

$$4\pi r^2 \rho(r) = 4\pi r^2 \rho_0 + \int_0^{Q_{\max}} QiQ \sin(Qr) \mathrm{d}Q \,, \tag{3.4}$$

Table 3.2. Structure parameters of a crystalline compound with $x = 0.91$ and an amorphous alloy with $x = 0.90$ for La$(Fe_{1-x}Al_x)_{13}$ [24, 33]

Pair	$x = 0.91$ (crystalline)		$x = 0.90$ (amorphous)	
	r [nm]	CN	r [nm]	CN
Fe–Fe	0.254	9.3	0.255 ± 0.002	9.2 ± 0.2
Fe–Fe	0.352	0.8	0.375 ± 0.005	1.1 ± 0.3
Fe–Fe	0.419	12.6	0.419 ± 0.002	10.8 ± 0.3
Fe–Fe	0.464	6.7	0.464 ± 0.002	7.3 ± 0.3
Fe–Fe	0.501	7.6	0.501 ± 0.002	7.7 ± 0.3
Fe–Fe	0.545	3.4	0.551 ± 0.002	4.8 ± 0.4
Fe–Fe	0.584	7.2	0.591 ± 0.002	6.6 ± 0.4
La–Fe	0.339	24.0	0.320 ± 0.002	17.2 ± 0.7
La–Fe	0.510	32.0	0.510 ± 0.002	26.9 ± 0.9
La–La	0.579	6.0	0.556 ± 0.003	5.2 ± 0.7

where $\rho(r)$ is the radial number density function, ρ_0 is the average number density of the sample and Q_{max} is the maximum scattering vector determined from the experimental condition. The RDFs for the La$(Fe_{1-x}Al_x)_{13}$ amor-

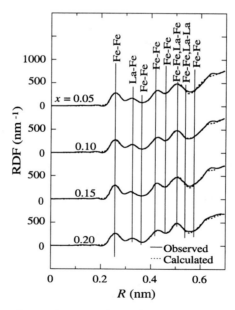

Fig. 3.7. Radial distribution functions (RDFs) of La$(Fe_{1-x}Al_x)_{13}$ amorphous alloys [24, 33]. *Solid line*: observation, *broken line*: calculated values

phous alloys as a function of distance R are depicted in Fig. 3.7. In Fig. 3.7, the solid and dotted curves denote the experimental and calculated RDFs obtained from Fourier transformation of the RIF evaluated from (3.2) and (3.3), respectively. The first peak composed of two peaks at about 0.25 and 0.32 nm is isolated from other peaks and fairly distinct oscillations are present even up to the middle distance range. These features also indicate the appearance of the SRO clusters in the $La(Fe_{1-x}Al_x)_{13}$ amorphous alloys.

Fe atoms occupy two different sites in the $La(Fe_{1-x}Al_x)_{13}$ compound, i.e., they are located at the center and vertices of the icosahedron (see Fig. 3.4c). The shortest Fe–Fe distance in the crystal is the distance between Fe^I and Fe^{II}. This distance is about 4% shorter than those between Fe^{II} and Fe^{II}. Although the result of the least-squares variational analysis for the amorphous alloy clearly suggests the appearance of icosahedral clusters in the amorphous alloy, the structural parameters are the average values for all Fe atoms in the icosahedral cluster. The structural parameters of the $La(Fe_{0.90}Al_{0.10})_{13}$ amorphous alloy are therefore compared with the average values calculated from the crystalline data summarized in Table 3.2. The distance of the nearest-neighbor Fe–Fe pair is about 0.255 nm. This value coincides with the Fe–Fe distance in the corresponding crystal. Furthermore, it is noteworthy that this distance hardly depends on the Fe concentration, as can be seen from Table 3.3 [24].

The coordination numbers of the nearest-neighbor La–Fe pairs provide us with structural information on the configuration of the icosahedral clusters around the La atom. These experimental results indicate that the distribution of the icosahedral clusters around the La atom in the amorphous alloy is extremely different from that in the crystal. It should therefore be emphasized that the structure of the amorphous alloys is not considered simply to be an assembly of microcrystals of $La(Fe_{1-x}Al_x)_{13}$. Figure 3.8 shows the concentration dependence of the CN of nearest-neighbor Fe–Fe pairs, together with that of the crystals [34]. The value decreases from 9.5 for $x = 0.05$ to 8.4 for $x = 0.20$, as can be seen from the figure, and the crystals exhibit almost the same values. Hence, the CN values of La–Fe pairs in the amorphous alloys are much smaller than those in the crystal. The change in the CN of the nearest-neighbor Fe–Fe pairs in these amorphous alloys is therefore attributed, not

Table 3.3. The nearest neighbor Fe–Fe distance r and the coordinate number CN for four kinds of $La(Fe_{1-x}Al_x)_{13}$ amorphous alloy [24]

x	r [nm]	CN
0.05	0.255 ± 0.002	9.5 ± 0.2
0.10	0.255 ± 0.002	9.2 ± 0.2
0.15	0.254 ± 0.002	8.3 ± 0.2
0.20	0.255 ± 0.002	8.4 ± 0.2

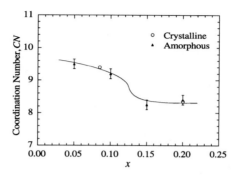

Fig. 3.8. Concentration dependence of the coordination number CN of the nearest-neighbor Fe–Fe pairs in the amorphous state [24], together with that in the crystalline state [34]

to any change in the internal structure of the icosahedral clusters, but rather to an increase in the Fe concentration. The coordination number of La–Fe^{II}, or La–Fe(Al) pairs, is smaller in the amorphous state than in the crystalline state, and the intra-atomic distance of the La–Fe^{II} nearest-neighbor pairs is shorter in the amorphous state than in the crystalline state. The La atom is totally surrounded by 8 icosahedra, i.e., 24 Fe atoms. From the ratio of the CN values for La–Fe(Al) pairs in the amorphous and crystalline states, it is found that the number of clusters around the La atom decreases to about three-quarters of that in the crystal (see Table 3.2). Thus, by dividing the CN of the La–Fe pairs in Table 3.2 by 3, it is found that the La atom is surrounded by about six icosahedral clusters. This number decreases with decreasing x. The average radius of the La atom in the amorphous alloys calculated from the nearest-neighbor Fe–Fe and La–Fe pairs in Table 3.2 is about 0.192 nm.

Since the icosahedral clusters in the amorphous alloy resemble those in the crystal, the radius of the icosahedral cluster in the amorphous alloy is calculated to be 0.289 nm from the lattice parameter of the crystal (0.577 nm) by assuming a sphere for the icosahedral cluster. The ratio of the radius of the La atom to that of the icosahedral cluster in the amorphous alloy is 0.665. These experimental results suggest that the La atom is located at the octahedral site formed by the icosahedral cluster. Consequently, it is clear that the atomic structure of the amorphous alloy is in striking contrast with that in the crystal, although the local structural unit is the icosahedral cluster of Fe^{II} atoms.

The RIFs $Qi(Q)$ and the radial distribution functions (RDFs) for RE(Fe$_{1-x}$Al$_x$)$_{13}$ (RE = Y, Ce and Lu) amorphous alloys are shown in Figs. 3.9 and 3.10, respectively. It should be noted that these systems form no NaZn$_{13}$-type compounds. The solid and broken curves express the experimental and calculated results, respectively. These results are very similar to those for the La(Fe$_{1-x}$Al$_x$)$_{13}$ amorphous alloys, implying the existence of icosahe-

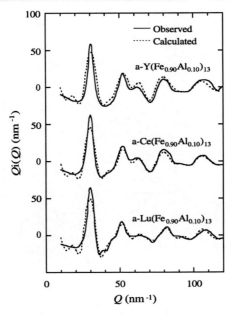

Fig. 3.9. Reduced interference functions $Qi(Q)$ of RE(Fe$_{0.90}$Al$_{0.10}$)$_{13}$ (RE = Y, Ce and Lu) amorphous alloys [35]. *Solid line*: observation, *broken line*: calculated values

dral clusters. As shown in Table 3.4, the Fe–Fe interatomic distance hardly depends on the kind of RE, but the RE–Fe distance is different, depending on the difference in their interatomic distance [20].

Table 3.4. Atomic structure parameters of RE(Fe$_{1-x}$Al$_x$)$_{13}$ amorphous alloys [20]

Pair	a-Ce(Fe$_{0.90}$Al$_{0.10}$)$_{13}$		a-Y(Fe$_{0.90}$Al$_{0.10}$)$_{13}$	
	r [nm]	CN	r [nm]	CN
Fe–Fe	0.254 ± 0.002	9.7 ± 0.2	0.254 ± 0.002	9.3 ± 0.2
Fe–Fe	0.360 ± 0.003	2.0 ± 0.2	0.378 ± 0.004	1.6 ± 0.3
Fe–Fe	0.421 ± 0.002	12.1 ± 0.3	0.420 ± 0.002	11.7 ± 0.3
Fe–Fe	0.465 ± 0.002	9.6 ± 0.3	0.465 ± 0.002	9.0 ± 0.3
Fe–Fe	0.502 ± 0.002	7.6 ± 0.3	0.500 ± 0.002	8.0 ± 0.4
Fe–Fe	0.545 ± 0.002	8.1 ± 0.4	0.548 ± 0.002	6.9 ± 0.4
Fe–Fe	0.590 ± 0.002	7.3 ± 0.4	0.591 ± 0.002	8.2 ± 0.9
RE–Fe	0.306 ± 0.002	18.9 ± 0.5	0.309 ± 0.002	18.1 ± 0.9
RE–Fe	0.505 ± 0.002	25.7 ± 0.9	0.513 ± 0.002	24.5 ± 1.6
RE–RE	0.553 ± 0.002	8.3 ± 0.8	0.554 ± 0.006	8.6 ± 2.2

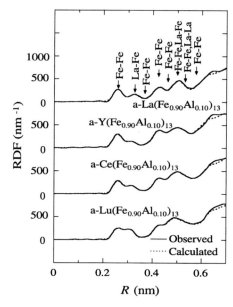

Fig. 3.10. Radial distribution functions (RDFs) of RE(Fe$_{0.90}$Al$_{0.10}$)$_{13}$ (RE = Y, Ce and Lu) amorphous alloys [35]. *Solid line*: observation, *broken line*: calculated values

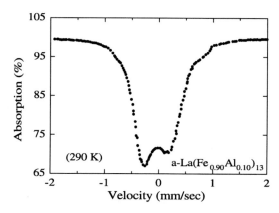

Fig. 3.11. Mössbauer spectra at 290 K of the La(Fe$_{0.90}$Al$_{0.10}$)$_{13}$ alloy in the amorphous state [24]

The existence of similar SRO in alloys can also be detected by Mössbauer spectroscopy for ^{57}Fe atoms. Indeed, the isomer shift (IS) and the quadrupole splitting (QS) of the paramagnetic Mössbauer spectra are sensitive to the s electron spatial density and local electric field gradients, respectively, which are modified by the local environment of the Fe nucleus. The paramagnetic Mössbauer spectra at 290 K for the La(Fe$_{0.90}$Al$_{0.10}$)$_{13}$ amorphous alloy and its

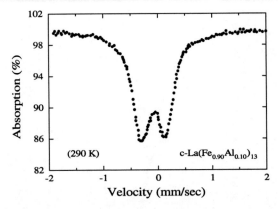

Fig. 3.12. Mössbauer spectra at 290 K of the La(Fe$_{0.90}$Al$_{0.10}$)$_{13}$ compound in the crystalline state [24]

crystalline counterpart are shown in Figs. 3.11 and 3.12, respectively [24]. The line width of the Mössbauer spectra is much wider than the natural line width of ^{57}Fe, which is ascribed to the distribution of the quadrupole interactions. The low-velocity absorption peak is slightly larger than the high-velocity peak. This asymmetric spectrum is attributed to the correlation between the distribution of IS and that of the quadrupole interactions. The average IS and QS of the La(Fe$_{0.90}$Al$_{0.10}$)$_{13}$ amorphous alloy and crystalline compound are summarized in Table 3.5, together with those of the La$_{17.5}$Fe$_{82.5}$ amorphous alloy for comparison [24, 35]. These two values for La(Fe$_{0.90}$Al$_{0.10}$)$_{13}$ are identical with those in the amorphous and crystalline states. This implies that the local environments around the Fe atoms in the two states are very similar to each other, consistent with the results obtained by X-ray diffraction. Furthermore, the average value of QS for the binary La$_{17.5}$Fe$_{82.5}$ amorphous alloy [37] is slightly larger than that for the La(Fe$_{0.90}$Al$_{0.10}$)$_{13}$

Table 3.5. The average isomer shift (IS) and the average quadrupole splitting (QS) of the La(Fe$_{0.90}$Al$_{0.10}$)$_{13}$ crystalline compound and several kinds of Fe-based amorphous alloys [24, 35]

Alloy	IS [mm/s]	QS [mm/s]
c-La(Fe$_{0.90}$Al$_{0.10}$)$_{13}$	−0.04	0.39
a-La(Fe$_{0.90}$Al$_{0.10}$)$_{13}$	−0.04	0.39
a-Ce(Fe$_{0.90}$Al$_{0.10}$)$_{13}$	−0.04	0.36
a-Y(Fe$_{0.90}$Al$_{0.10}$)$_{13}$	−0.04	0.36
a-Ce$_{15}$Fe$_{85}$	−0.08	0.40
a-La$_{17.5}$Fe$_{82.5}$	−0.06	0.44

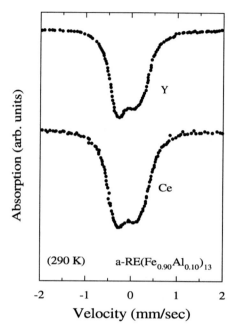

Fig. 3.13. Mössbauer spectra at 290 K of Y(Fe$_{0.90}$Al$_{0.10}$)$_{13}$ and Ce(Fe$_{0.90}$Al$_{0.10}$)$_{13}$ amorphous alloys [35]

amorphous alloy, although the Fe content in both alloys is not so different. This suggests that there is some difference between the local atomic arrangements in La(Fe$_{0.90}$Al$_{0.10}$)$_{13}$ and La$_{17.5}$Fe$_{82.5}$ amorphous alloys. That is, in the former amorphous alloy, the presence of icosahedral clusters would suppress fluctuations in the local atomic arrangements, such as the number of nearest-neighbor atoms and the interatomic distance.

The Mössbauer spectra at 290 K for the Y(Fe$_{0.90}$Al$_{0.10}$)$_{13}$ and Ce(Fe$_{0.90}$Al$_{0.10}$)$_{13}$ amorphous alloys in the paramagnetic state are shown in Fig. 3.13. The average IS and QS of Y(Fe$_{0.90}$Al$_{0.10}$)$_{13}$ and Ce(Fe$_{0.90}$Al$_{0.10}$)$_{13}$ amorphous alloys are also supplied in Table 3.5 [38]. The average IS values for RE(Fe$_{0.90}$Al$_{0.10}$)$_{13}$ amorphous alloys (RE = La, Ce and Y) coincide with one another, suggesting a similar electronic structure. A slight difference in QS among RE(Fe$_{0.90}$Al$_{0.10}$)$_{13}$ amorphous alloys probably arises from the difference in their atomic size. In the same table, the data for the Ce$_{15}$Fe$_{85}$ binary amorphous alloy are also contrasted with those for the Ce(Fe$_{0.90}$Al$_{0.10}$)$_{13}$ amorphous alloy described above.

3.2.3 La(Ni$_{1-x}$Al$_x$)$_{13}$ Amorphous Alloys

In crystalline compounds, not only La(Fe$_{1-x}$Al$_x$)$_{13}$ ($0.08 \leq x \leq 0.54$) but also La(Co$_{1-x}$Al$_x$)$_{13}$ ($0 \leq x \leq 0.3$) and La(Ni$_{1-x}$Al$_x$)$_{13}$ ($0.31 \leq x \leq 0.38$), the

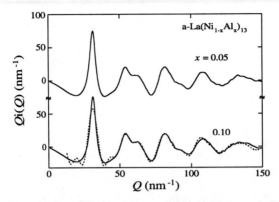

Fig. 3.14. Reduced interference functions $Qi(Q)$ of La(Ni$_{1-x}$Al$_x$)$_{13}$ amorphous alloys with $x = 0.05$ and 0.10 [41]. *Solid line*: observation, *broken line*: calculated values

systems have the NaZn$_{13}$-type structure [39, 40]. Therefore, similar clusters are expected to form in the amorphous counterparts.

Figure 3.14 represents the RIFs $Qi(Q)$ of the La(Ni$_{1-x}$Al$_x$)$_{13}$ ($x = 0.05$ and 0.10) amorphous alloys obtained by X-ray diffraction. The significant feature of $Qi(Q)$ is a long-period oscillation persisting up to high-Q regions, in analogy with the La(Fe$_{1-x}$Al$_x$)$_{13}$ amorphous alloys. Such a profile reveals the presence of well-defined SRO clusters, as mentioned in Sect. 3.2.2. Furthermore, the profiles for $x = 0.05$ and 0.10 are quite similar to each other, revealing the existence of the same short-range clusters. In the fitting procedure, r and CN values of near-neighbor correlations of a NaZn$_{13}$-type crystalline compound are adopted as the starting parameters and the resultant $Qi(Q)$ of La(Ni$_{0.90}$Al$_{0.10}$)$_{13}$ is plotted by the dashed line in Fig. 3.14. The oscillation

Table 3.6. Structural parameters, the interatomic distance r and coordination number CN, for the La(Ni$_{0.90}$Al$_{0.10}$)$_{13}$ amorphous alloy [41]

Pair	r [nm]	CN
Ni–Ni	0.247	8.4
Ni–Ni	0.381	1.3
Ni–Ni	0.417	9.8
Ni–Ni	0.463	8.1
Ni–Ni	0.501	7.1
Ni–Ni	0.550	3.7
Ni–Ni	0.594	9.1
La–Ni	0.302	15.5
La–Ni	0.497	16.9
La–La	0.547	7.5

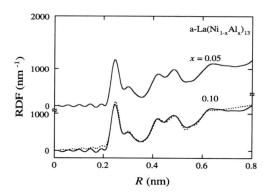

Fig. 3.15. Radial distribution functions (RDFs) of La(Ni$_{1-x}$Al$_x$)$_{13}$ amorphous alloys with $x = 0.05$ and 0.10 [41]. *Solid line*: observation, *broken line*: calculated values

of $Qi(Q)$ with a long period is well represented without a large deviation of structural parameters from those of the crystalline compounds. The parameters obtained are listed in Table 3.6. Considering the size difference between Ni and Fe atoms, r and CN values of the Ni–Ni pair make it clear that icosahedral clusters also exist in the La(Ni$_{1-x}$Al$_x$)$_{13}$ amorphous alloys. The distance of the La–La pair in the amorphous state is 0.547 nm, comparable with half the lattice constant of the La(Ni$_{1-x}$Al$_x$)$_{13}$ crystalline compounds. As a result, the La atoms are separated from each other by the icosahedral clusters. The coordination number of La–Ni correlation in the amorphous state is much smaller than that of the NaZn$_{13}$-type compound, indicating that the distribution of icosahedral clusters around La atoms differs from that in the NaZn$_{13}$-type crystalline state, in analogy with La(Fe$_{1-x}$Al$_x$)$_{13}$ amorphous alloys.

The RDFs obtained from $Qi(Q)$ for the La(Ni$_{1-x}$Al$_x$)$_{13}$ amorphous alloys with $x = 0.05$ and 0.10 are displayed by the solid line in Fig. 3.15, together with the calculated curve transformed from fitted results on $Qi(Q)$ for $x = 0.10$ (broken line). Very similar oscillations are observed in the experimental curves for $x = 0.10$ and 0.05 up to around 0.60 nm. Such a striking resemblance is due to the existence of icosahedral clusters. The RDF obtained from the calculated $Qi(Q)$, indicated by the broken line, fully represents the experimental one, revealing that the distinct oscillations in $Qi(Q)$ originate in icosahedral clusters.

La(Ni$_{1-x}$Al$_x$)$_{13}$ crystalline compounds are formed in the concentration range $0.31 \leq x \leq 0.38$, and no stable compound is confirmed in the range $x < 0.31$. Icosahedral clusters in amorphous alloys are formed over much wider concentration ranges. Accordingly, the onset of ferromagnetism is expected in the amorphous alloys with $0.05 \leq x \leq 0.10$, in contrast to the Pauli paramagnetic La(Ni$_{1-x}$Al$_x$)$_{13}$ ($0.31 \leq x \leq 0.38$) crystalline compounds. Furthermore, the ferromagnetic properties are controlled by changing the Ni

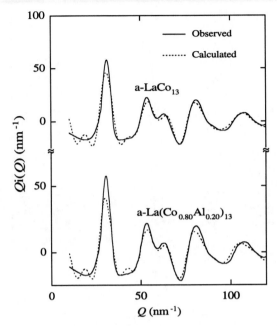

Fig. 3.16. Reduced interference functions $Qi(Q)$ of LaCo$_{13}$ and La(Co$_{0.80}$Al$_{0.20}$)$_{13}$ amorphous alloys [42]. *Solid line*: observation, *broken line*: calculated values

concentration whilst keeping the icosahedral-cluster-type SRO. Consequently, the change in the SF features can be discussed by neglecting the change in the degree of SRO due to the change in the concentration.

3.2.4 La(Co$_{1-x}$Al$_x$)$_{13}$ and La(Co$_{1-x}$Mn$_x$)$_{13}$ Amorphous Alloys

LaCo$_{13}$ is the only binary compound with the NaZn$_{13}$-type structure amongst RE–TM systems. Since LaCo$_{13}$ contains no additional third element, it is useful in shedding light on the local atomic arrangement in the amorphous alloy and on the difference between the magnetic properties in the amorphous and crystalline states.

The RIFs $Qi(Q)$ of LaCo$_{13}$ and La(Co$_{0.80}$Al$_{0.20}$)$_{13}$ amorphous alloys are given in Fig. 3.16 [42]. The interatomic distance r and coordination number CN of near-neighbor correlations are obtained by the least-squares calculation so as to reproduce the experimental interference function data in the same way as for the La(Fe$_{1-x}$Al$_x$)$_{13}$ and La(Ni$_{1-x}$Al$_x$)$_{13}$ amorphous alloys. The broken lines represent calculated results. The starting parameters for the least-squares variations were set by calculating the average CN and r values at 300 K for the LaCo$_{13}$ crystal. Since the r and CN values of the nearest-neighbor Co–Co pairs for the amorphous alloy and for the crystalline compound show little difference, it is also plausible that local SRO clusters

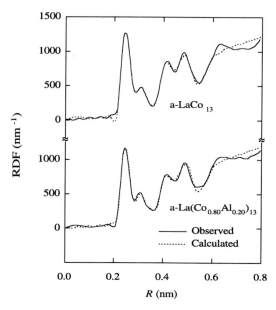

Fig. 3.17. Radial distribution functions (RDFs) of $LaCo_{13}$ and $La(Co_{0.80}Al_{0.20})_{13}$ amorphous alloys [42]. *Solid line*: observation, *broken line*: calculated values

in the amorphous state are icosahedral clusters, in analogy with the situation in the crystalline state [18, 19]. The essential features in the structural profiles are fairly distinct oscillations observed even in the high-Q region, being similar to those for oxide glasses [27], $La(Fe_{1-x}Al_x)_{13}$ and $La(Ni_{1-x}Al_x)_{13}$ amorphous alloys. Accordingly, the characteristic profiles also imply the formation of chemical SRO icosahedral clusters with a definite relation between the bond lengths and the angular relations [26].

Figure 3.17 shows the RDFs of the $LaCo_{13}$ and $La(Co_{0.80}Al_{0.20})_{13}$ amorphous alloys [42]. The solid and broken curves were obtained by Fourier transformation of the experimental and calculated RIFs in Fig. 3.16. The first peak, composed of two peaks, is isolated from other peaks and fairly distinct oscillations are present even up to the middle-distance range. These features also indicate the formation of SRO clusters in the $LaCo_{13}$ amorphous alloy. Co atoms occupy two different sites, i.e., Co^{I} and Co^{II}. The shortest Co–Co distance is the distance between the Co^{I} and Co^{II}, being several percent shorter than the Co^{II}–Co^{II} distance. The structural parameters determined in the analyses are the average values for Co^{I} and Co^{II}, although the result of the least-squares variational analysis for the amorphous alloy clearly suggests the presence of icosahedral clusters. In Table 3.7, the nearest-neighbor interatomic distances r of the Co–Co and La–Co pairs are 0.250 nm and 0.311 ± 0.001 nm, respectively. The coordination numbers CN for these two pairs are 10.0 ± 0.2 and 18.7 ± 0.9, respectively. The structure parameters r

Table 3.7. Structural parameters of the LaCo$_{13}$ crystalline compound (c-LaCo$_{13}$) and its amorphous counterpart (a-LaCo$_{13}$), and the La(Co$_{0.95}$Mn$_{0.05}$)$_{13}$ amorphous alloy [42, 44]

Pair	c-LaCo$_{13}$		a-LaCo$_{13}$		a-La(Co$_{0.95}$Mn$_{0.05}$)$_{13}$	
	r [nm]	CN	r [nm]	CN	r [nm]	CN
Co–Co	0.251	10.2	0.248	9.3	0.248	9.1
Co–Co	0.333	0.9	0.371	1.7	0.363	1.7
Co–Co	0.411	13.9	0.417	16.4	0.416	16.1
Co–Co	0.456	9.2	0.466	9.7	0.464	9.4
Co–Co	0.510	10.2	0.523	10.3	0.523	11.0
Co–Co	0.568	7.9	0.556	5.3	0.554	5.5
Co–Co	0.599	14.8	0.599	7.9	0.599	7.8
La–Co	0.329	24.0	0.309	24.3	0.310	23.7
La–Co	0.509	32.0	0.492	29.8	0.492	28.2
La–La	0.567	6.0	0.561	5.1	0.564	5.7

and CN for the Co–Co pairs in the amorphous state are very similar to those in the crystalline state, whereas those for La–Co pairs are strikingly different, as can be seen from the table.

The RDF of the La(Co$_{0.95}$Mn$_{0.05}$)$_{13}$ amorphous alloy is given in Fig. 3.18. Solid lines stand for experimental results and broken lines indicate the calculated RDF obtained by fitting to the experimental RIF, using the structure parameters of the LaCo$_{13}$ crystalline compound as starting parameters. It has been confirmed by low temperature XAFS spectra that Mn atoms occupy the Co^{II} sites of La(Co$_{0.95}$Mn$_{0.05}$)$_{13}$ in the crystalline state [45]. The

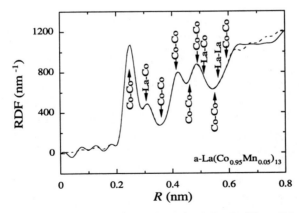

Fig. 3.18. Radial distribution function (RDF) of the La(Co$_{0.95}$Mn$_{0.05}$)$_{13}$ amorphous alloy [81]. *Solid line*: observation, *broken line*: calculated values

RDF of this ternary amorphous alloy is very similar to the RDF of the $LaCo_{13}$ binary amorphous alloy. The nearest-neighbor r and CN of the $LaCo_{13}$ amorphous alloy are nearly equal to those of the crystalline counterpart and no directly neighboring La–La pair exists in the experimental RDF, indicating that icosahedral clusters separate La atoms in the amorphous alloy [42]. As a consequence, the $La(Co_{0.95}Mn_{0.05})_{13}$ ternary amorphous alloy also has SRO defined by icosahedral clusters arranged randomly around La atoms. We will now leave the atomic structures and turn to the magnetic properties of $RE(TM_{1-x}Al_x)_{13}$ amorphous alloys.

3.3 Theoretical Aspects of Magnetic Excitations and Spin Fluctuations in Itinerant-Electron Ferromagnets

Compared with the localized magnetic moment systems, the elementary excitations are very diverse in itinerant-electron ferromagnetic systems, because $3d$ electrons move in the bulk under the influence of their mutual Coulomb interaction [2, 4, 43, 46, 47]. A unified explanation for all magnetic excitations at finite temperatures has not yet been established, because magnetic excitations in itinerant-electron magnetic systems involve complex interactions among $3d$ electrons. However, it has been recognized that some kinds of excitation, such as the spin-wave excitations, can be explained in detail in WIE ferromagnets. In this section, we review theoretical aspects of WIE ferromagnets for single-particle excitations, spin-wave excitations and SFs with mode–mode coupling.

3.3.1 Single-Particle Excitations

Magnetic properties of crystalline $3d$ TMs and alloys have been explained in terms of the energy band models. It is well known that the Stoner model on the basis of the Hartree–Fock (HF) approximation [3, 48] is a useful theory for describing the ground state of itinerant-electron ferromagnets. However, the elementary excitation at finite temperatures is limited to a single-particle excitation called the Stoner excitation. As a consequence, the Curie temperature T_C calculated from this theory is much higher than the experimental value, and the temperature dependence of paramagnetic susceptibility differs from the experimental results, which exhibit a Curie–Weiss behavior. It is well known that the Stoner excitation exhibits a variation of magnetization M as a function of temperature T in the form M–T^2. For WIE ferromagnets in which the exchange split of the $3d$ band is small, the magnetic equation of state can be expanded as an even power series in M. By neglecting terms higher than M^2, the relation between M and T is expressed in the form M^2–T^2.

3.3.2 Collective Excitations

In the Stoner theory, holes are formed by excitation of electrons to a higher energy state with different spin direction, and excited electrons and formed holes move independently in the common mean field. The importance of the interaction between the electrons and holes was first pointed out by Herring and Kittel (HK) [49, 50] and an improved treatment with a dynamical HF approximation was developed [51]. This approximation is referred to as the random phase approximation (RPA). The RPA gives important results on the spin-wave excitation generally observed in itinerant-electron ferromagnets. According to the RPA, the spin-wave dispersion coefficient D is given by

$$\hbar\omega = Dq^2 , \tag{3.5}$$

with

$$D = \frac{\hbar^2}{2m}\frac{1}{\zeta}\left[1 - \frac{3}{2}\frac{1}{I\rho(\varepsilon_F)}\frac{(1+\zeta)^{5/3} - (1-\zeta)^{5/3}}{\zeta}\right]$$

and

$$\zeta = \frac{M}{N_e} ,$$

where m, $\rho(\varepsilon_F)$, N_e and I are the mass of the electron, the DOS at the Fermi level, the total number of electrons and the Coulomb interaction per electron, respectively. The temperature dependence of magnetization due to the spin-wave excitation is given in the following form, in a similar manner to the localized magnetic moment:

$$M(T) = M(0)(1 - B_M T^{3/2}) , \tag{3.6}$$

with

$$B_M = 2.612\frac{g\mu_B}{M(0)}\left(\frac{k_B}{4\pi D}\right)^{3/2} .$$

The $T^{3/2}$ dependence of magnetization is observed in itinerant-electron ferromagnets, and the magnitude of D is consistent with the experimental result by neutron scattering [52]. Thus, the spin-wave excitations in itinerant-electron ferromagnets at finite temperatures can be discussed by the RPA. However, the value of the Curie temperature and the Curie–Weiss behavior cannot be explained, because of insufficient treatment of the mode–mode coupling of SFs in an equilibrium state at finite temperatures [47].

Since the order parameter of the ferromagnetic phase is the magnetization, the fluctuations of the local spin density are coupled with the free energy of the magnetic state at finite temperatures [46, 53]. As a result, the thermal equilibrium state is influenced by SFs and dynamical properties of SFs are determined by the thermal equilibrium state through the fluctuation–dissipation relation. Therefore, SFs are connected self-consistently with the free energy [46,53]. If only long-wavelength SFs are excited, SFs and the free energy can be treated theoretically by a linear response relation [54]. Moriya and Kawabata (MK) established a theory starting from this self-consistent relation [46,53]. The main results of their theory are expressed as follows:

$$F(T) = F_{\mathrm{HF}}(T) + F_{\mathrm{sf}}(T) , \tag{3.7}$$

with

$$F_{\mathrm{sf}}(T) = \frac{1}{N_0} \sum_q \int_0^\infty \frac{f(\omega)}{\pi} \mathrm{Im}\chi^{+-}(q,\omega)\mathrm{d}\omega$$

and

$$f(\omega) = \frac{\omega}{2} + T\ln\left[1 - \exp\left(-\frac{\omega}{T}\right)\right] ,$$

where $F(T)$ is the free energy, F_{HF} and F_{sf} are respectively the classical HF term and the correcting term due to SFs, and N_0 represents the total number of magnetic atoms. In more detail, the dynamical susceptibility with respect to the frequency ω and the wavenumber q under the magnetization M and the interaction I satisfies the self-consistent relation

$$\chi_{M,I}^{+-}(q,\omega) = \frac{\chi_{M,0}^{+-}(q,\omega)}{1 - I\chi_{M,0}^{+-}(q,\omega) + \lambda} . \tag{3.8}$$

The parameter λ includes the influence of interaction between SFs with different modes, which is the so-called mode–mode coupling, being neglected in the RPA. This self-consistent renormalization (SCR) treatment for the thermodynamic properties of the magnetic phase successfully explains the many characteristics of magnetic properties, in particular, those of WIE ferromagnets [2].

It is noteworthy that this theory can explain a Curie–Weiss law for WIE ferromagnets [53]. To be precise, the Curie–Weiss law for itinerant-electron systems is due to the temperature-dependent fluctuation amplitude of the local spin density, in contrast to the conventional Curie–Weiss law originating in the directional fluctuation of the localized magnetic moment with a constant amplitude. Consequently, the effective moment P_{eff} no longer has a correlation with the spontaneous moment P_{S}, giving a large ratio of $P_{\mathrm{eff}}/P_{\mathrm{S}}$. This ratio has been used as a measure of electron itinerancy in magnetic materials, because it should be unity in localized moment magnets. It is also known that the ratio is correlated to T_{C}. The $P_{\mathrm{eff}}/P_{\mathrm{S}}$ vs. T_{C} plots are called Rhodes–Wohlfarth (RW) plots [55]. The distribution of data plotted around a universal line was recognized empirically, whereas the physical meaning remained an open question for a long time. The SCR theory gives a physical meaning to the RW plots. To put it precisely, the longitudinal SFs become dominant as the magnetic properties come close to those of WIE ferromagnets and the inverse magnetic susceptibility is governed by the amplitude fluctuation rather than the directional fluctuation [2]. Furthermore, the temperature dependence of the spontaneous magnetization (M_{SP}) for WIE ferromagnets is also explained by the SCR theory.

In the SCR theory, the square of the magnetization M^2 is related to the temperature dependence of the correcting term of the free energy F_{sf}, and given by [2, 43]

$$M^2 \propto c_1 - \frac{1}{M}\frac{\partial F_{sf}}{\partial M} \propto c_1 - \int_0^\infty \frac{K(\omega)}{e^{\omega/k_B T} - 1} d\omega , \qquad (3.9)$$

with

$$K(\omega) = c_2\omega^{1/3}\left[\tan^{-1}\left(\frac{2q_0^2 - \omega^{2/3}}{\omega^{2/3}}\right) + c_3 \log\left(\frac{(q_0^2 + 3\pi\omega^{2/3})^3}{q_0^6 + 3\pi\omega^2}\right)\right]$$
$$+ c_4\frac{q_0^4\omega}{q_0^6 + 3\pi\omega^2} ,$$

where q_0 is the boundary wavenumber of the real spin-wave in paramagnon-like SF regimes. Since the frequency ω corresponds to the energy of excited fluctuations, the temperature dependence of M^2 is governed by the leading term in $K(\omega)$ at various temperatures. The calculated results indicate that M^2 is proportional to T^2 at temperatures far below T_C because $K(\omega) \propto \omega$, and to $T^{4/3}$ in the vicinity of T_C because $K(\omega) \propto \omega^{1/3}$ [43]. In particular, the latter dependence is characteristic of this theory, distinguishing it from other theories. In fact, this relation has been observed in WIE ferromagnets such as Sc_3In [56] and Ni_3Al [57].

3.3.3 Spin Fluctuations at Finite Temperatures and Thermal Expansion Anomaly

From investigations of various WIE ferromagnets, it has been revealed that SF features can be classified with respect to the flexibility in the amplitude of local moments and the magnitude of localization of fluctuations in the wavenumber and in the real spaces [2]. This is shown in Fig. 3.19. The abscissa indicates the magnitude of localization and the ordinate shows the flexibility of SF amplitudes. The temperature dependence of SF amplitudes is connected with the dynamical susceptibility $\chi^{+-}(q,\omega)$ in (3.8) and is expressed by

$$S_L^2(T) = \frac{3}{\pi N_0^2}\sum_q \int_0^\infty n(\omega)\ln\chi^{+-}(q,\omega)d\omega , \qquad (3.10)$$

with

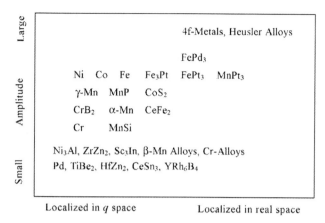

Fig. 3.19. Classification of spin fluctuation (SF) features for various crystalline ferromagnetic and antiferromagnetic materials [2]

$$n(\omega) = \frac{\exp(-\omega/T)}{1 - \exp(-\omega/T)}.$$

In Fig. 3.19, the left-hand side is the WIE ferromagnet limit where $\mathrm{Im}\chi^{+-}(q,\omega)$ has spectral intensity only around $q \approx 0$. In other words, the SFs localize in the wavenumber space. On the other hand, the right-hand side is the localized magnetic moment limit where $\mathrm{Im}\chi^{+-}(q,\omega)$ becomes a broad spectrum up to high q and SFs localize in the real space. SF amplitudes are small and depend on temperature on the lower side. The amplitude becomes larger as we shift to the upper side, and eventually the amplitude is fixed and independent of temperature. For instance, WIE ferromagnets such as $\mathrm{Sc_3In}$ and $\mathrm{Ni_3Al}$ exhibit a narrow spectral width of $\mathrm{Im}\chi^{+-}(q,\omega)$ around $q \approx 0$ and a small $S_{\mathrm{L}}^2(T)$ which depends on temperature. In $4f$ metals, the magnetic $4f$ electrons localize on the atomic site, and hence $\mathrm{Im}\chi^{+-}(q,\omega)$ is characterized by the periodicity of $4f$ atoms. The amplitude of $S_{\mathrm{L}}^2(T)$ in $4f$ metals is influenced by mainly intra-atomic exchange interactions and becomes large but exhibits no temperature dependence. As can be seen from Fig. 3.19, many magnetic substances exhibit intermediate values of the spectral width of $\mathrm{Im}\chi^{+-}(q,\omega)$ and the magnitude of thermal variation of $S_{\mathrm{L}}^2(T)$. The $3d$ metals, Fe, Co and Ni, are itinerant-electron magnets with a relatively large $S_{\mathrm{L}}^2(0)$ in the ground state and weak temperature dependence of $S_{\mathrm{L}}^2(T)$ at finite temperatures. Alloys and compounds consisting of these $3d$ metals show a different width of $\mathrm{Im}\chi^{+-}(q,\omega)$ together with a different magnitude of thermal variation of $S_{\mathrm{L}}^2(T)$ with respect to the atomic and electronic structures.

The local spin density also fluctuates in itinerant-electron amorphous ferromagnets. This is evidence that the SF amplitude varies with temperature and SF features are different with respect to $3d$ elements. Nevertheless, few detailed investigations on SFs have been carried out for amorphous materials, except for some results specifying the 4/3 power of the temperature dependence of the square of the magnetization [58, 59]. Checking the validity of the M^2–$T^{4/3}$ relation is a straightforward analysis in crystalline WIE ferromagnets [56, 57] and this relation is theoretically supported for homogeneous systems [43]. However, it is difficult, even in crystalline materials, to determine whether this relation is held exactly or not, because it is only valid over a limited temperature range around T_C [43, 46, 53]. Therefore, much further information is needed to reveal the SF features in amorphous ferromagnets.

3.4 Fundamental Magnetic Properties of La(TM$_{1-x}$Al$_x$)$_{13}$ and La(Co$_{1-x}$Mn$_x$)$_{13}$ Amorphous Alloys

Magnetic properties are quite sensitive to atomic structures. In this section, various magnetic properties of La(Ni$_{1-x}$Al$_x$)$_{13}$, La(Co$_{1-x}$Al$_x$)$_{13}$, La(Co$_{1-x}$Mn$_x$)$_{13}$ and La(Fe$_{1-x}$Al$_x$)$_{13}$ amorphous alloys are discussed in connection with atomic structures and SFs in the amorphous state. Recently, an attempt has been made to map the SF features in Y–TM (TM = Mn, Fe, Co and Ni) amorphous alloys by comparing the temperature dependence of magnetization, the Arrott plot and the thermal expansion [60]. The results classified using these data are mapped in Fig. 3.20 [61] in accordance with

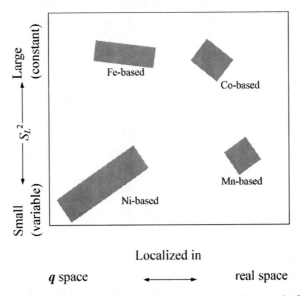

Fig. 3.20. Classification of spin fluctuation features for Fe-based, Co-based, Ni-based and Mn-based amorphous alloys [60]

Fig. 3.19. As can be seen from the figure, the regions for these TM are clearly divided, indicating that SF features are well-defined one by one. From the figure, it should be noted that Ni-based amorphous alloys have small amplitude local moment m and localized SFs in the wavenumber space. In such a regime, longitudinal SFs are the dominant elementary excitations at finite temperatures [2,43,46,53,62].

In Fe- and Mn-based amorphous alloys, spin-glass behavior is observed, indicating that the spatial fluctuation of the magnetic moment is dominant. However, SFs in these amorphous alloys are very different from each other as regards the magnetovolume effect. Putting it another way, a large spontaneous volume magnetostriction is observed in Fe-based amorphous alloys, while Mn-based amorphous alloys exhibit a very small magnetovolume effect, although the magnetic moment in the ground state of Mn-based alloys is larger than that in Fe-based alloys. In Co-based amorphous alloys, the spectral width of $\mathrm{Im}\chi^{+-}(q,\omega)$ evaluated from the Arrott plot shows that the SF is relatively localized in real space. Consequently, it is expected that the SF features in the amorphous state can be verified and compared with those in the crystalline state.

3.4.1 La(Ni$_{1-x}$Al$_x$)$_{13}$ Amorphous Alloys

Weak itinerant-electron (WIE) ferromagnets are considered as good candidates for detailed discussion, not only of the mechanism underlying the onset of itinerant-electron ferromagnetism, but also of magnetism at finite temperatures [2]. Among many transition-metal-based amorphous alloys, Ni-based alloys tend to exhibit WIE ferromagnetic properties, and investigations of magnetic properties for Ni-based amorphous alloys have been carried out. However, the discussion has focused on only a limited range of SF features [14–16,63].

In binary amorphous alloy systems, very careful attention has been paid to the itinerant-electron ferromagnetism of Ni$_{81.6}$B$_{18.4}$ [15] and Y$_x$Ni$_{100-x}$ ($3.0 \leq x \leq 12.8$) [16] amorphous alloys, which have been discussed in terms of the Edwards–Wohlfarth (EW) model [63]. This model is based on the Stoner theory and takes only the single-particle excitation into account. According to the EW model, the temperature (T) and magnetic field (H) dependences of the magnetization (M) are expressed in the following way [64]:

$$M^2(H,T) = M^2(0,0)\left[1 - \left(\frac{T}{T_C}\right)^2 + 2\chi_0\frac{H}{M(H,T)}\right], \qquad (3.11)$$

where T_C is the Curie temperature and χ_0 is the zero-field susceptibility. Therefore, M_{SP} varies as a function of $M(0,T)^2 - T^2$. In addition, (3.11) explains the linear variation of the Arrott plots. The temperature T dependence of magnetization M for Ni$_{81.6}$B$_{18.4}$ follows the M–$T^{3/2}$ relation at low temperatures, indicating a spin-wave contribution to the thermomagne-tization curve. On the other hand, an M–T^2 relation appears with increas-ing temperature. This M–T^2 relation has been considered to be due to the single-particle excitation as expected from the EW model. Note that the M–T^2 relation with a small M is equal to the M^2–T^2 relation since the M^4 term is negligibly small. Since the expected Curie temperature from the EW model is determined by the single-particle excitations [63], the M–T^2 relation should hold up to T_C. However, it has been pointed out that the curve for the Ni$_{81.6}$B$_{18.4}$ amorphous alloy deviates from the M–T^2 relation with increasing temperature up to T_C [15].

Similar analyses have been performed for the M–T curves of Y–Ni amor-phous alloys with various Ni concentrations [16]. In this system, the M^2–T^2 relation is clearly observed over the whole temperature range below the Curie temperature in the high Ni concentration range. On the other hand, the rela-tion becomes invalid with decreasing T_C, or with decreasing Ni concentration. Liénard and Rebouillat have pointed out that the contribution from collective excitations is overlooked in the conventional theories [16]. In addition, it has been noted that magnetic inhomogeneity in the amorphous state explains the discrepancies between these results in the EW model [65, 66]. Actually, mag-netic properties of Ni–P amorphous alloys indicate the existence of magnetic clusters with decreasing Ni concentration [63]. Furthermore, a blocking of isolated magnetic clusters has been confirmed in the Y$_{23.7}$Ni$_{76.3}$ amorphous alloy with a slightly lower Ni concentration than the critical concentration at the onset of ferromagnetism [67].

The above argument seems to provide a sufficient explanation for peculiar magnetic properties of Ni-based amorphous alloys. However, one should re-member that the deviation of the temperature dependence of magnetization from the EW model for the Ni$_{81.6}$B$_{18.4}$ amorphous alloy has been observed even though the Arrott plots exhibit a clear linearity [15]. This linearity has also been confirmed in Y–Ni amorphous alloys in the concentration range where the M^2–T^2 relation is invalid [16]. Accordingly, collective SFs are ex-pected to have some effect on the magnetic properties of the Y–Ni amorphous alloys.

Figure 3.21 shows magnetization curves at 4.2 K for the La(Ni$_{1-x}$Al$_x$)$_{13}$ amorphous alloys. For the highest Ni concentration alloy, the magnetization is almost saturated above 5 kOe. On the other hand, the slope after satu-ration or the high-field susceptibility χ_{hf} become larger with increasing x. In particular, χ_{hf} of the La(Ni$_{0.90}$Al$_{0.10}$)$_{13}$ amorphous alloy is very large, as can be seen from Fig. 3.21. The magnetization process of WIE ferromagnets has an interrelation with the spectral width of SFs. Consequently, changes in the features of the magnetization curves should be related to changes in SF features.

The spontaneous magnetic moment M_{Ni} per Ni atom for La(Ni$_{1-x}$Al$_x$)$_{13}$ amorphous alloys is plotted against the Ni concentration in Fig. 3.22a. In

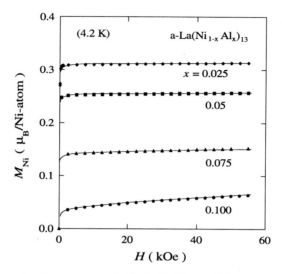

Fig. 3.21. Magnetization curves at 4.2 K for $La(Ni_{1-x}Al_x)_{13}$ amorphous alloys [61]

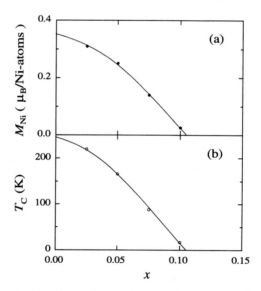

Fig. 3.22. Concentration dependence of magnetic properties for $La(Ni_{1-x}Al_x)_{13}$ amorphous alloys [61]. (**a**) Magnetic moment M_{Ni} per Ni atom, (**b**) Curie temperature T_C

the vicinity of the critical concentration for the onset of ferromagnetism, the spontaneous magnetic moment M_{Ni} decreases linearly with increasing x, whereas the concentration dependence of M_{Ni} becomes sluggish in higher Ni concentration ranges. The value of M_{Ni} extrapolated to $x = 0$ is 0.35–0.40μ_B. The Curie temperature T_C is determined from the Arrott plots at

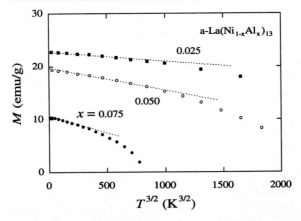

Fig. 3.23. Temperature dependence of magnetization M in the form M–$T^{3/2}$ for La(Ni$_{1-x}$Al$_x$)$_{13}$ amorphous alloys [41]

various temperatures and the concentration dependence of T_C is displayed in Fig. 3.22b. The value of T_C for all the specimens lies below room temperature and features in the concentration dependence of T_C are similar to those of M_{Ni}. It is extrapolated to be 240 K at $x = 0$.

Temperature Dependence of Magnetization
Below the Curie Temperature

For the La(Ni$_{1-x}$Al$_x$)$_{13}$ amorphous alloys, the change in the local atomic environment with concentration can be neglected, because of the formation of icosahedral clusters with well-defined bond lengths and angular relations. Therefore the elementary excitation properties as well as the stability of the ferromagnetic phase can be investigated in detail over a wide concentration range. In this section, we discuss the influence of spin waves, SFs with mode–mode coupling and single-particle excitations on the thermomagnetization curves.

Even in itinerant-electron ferromagnets, spin-wave excitations occur and contribute to the thermal variation of magnetization in the form of M–$T^{3/2}$ at low temperatures [49, 50]. The $T^{3/2}$ dependence is observed at low temperatures in La(Ni$_{0.925}$Al$_{0.075}$)$_{13}$, La(Ni$_{0.95}$Al$_{0.05}$)$_{13}$ and La(Ni$_{0.975}$Al$_{0.025}$)$_{13}$ amorphous alloys, as shown in Fig. 3.23. The valid range is about 0.15–0.17T_C for the former two amorphous alloys and about 0.3T_C for the latter. In the La(Ni$_{0.90}$Al$_{0.10}$)$_{13}$ amorphous alloy, it is difficult to ascertain the $T^{3/2}$ dependence. It has been pointed out that the M–$T^{3/2}$ relation is valid only in low temperature regions in the WIE ferromagnet limit [2], because the interaction between spin waves becomes significant with increasing temperature. The valid temperature range for the $T^{3/2}$ relation thus depends on the number of excited spin waves. In other words, as the excitation probability becomes high, the $T^{3/2}$ relation only holds for very low temperatures. Another reason for the invalidity of the $T^{3/2}$ relation is the existence of other excitations such as single-particle excitations. For the La(Ni$_{1-x}$Al$_x$)$_{13}$ amorphous alloys, the validity range of the $T^{3/2}$ relation becomes narrower with decreasing T_C. This implies that contributions from paramagnon-like spin-wave excitations and single-particle excitations to the magnetic properties vary with the Ni concentration.

The spin-wave dispersion coefficient D gives further important information. For spin waves in the ferromagnetic state, the dispersion relation between the wavenumber q and the frequency ω is given by $\hbar\omega = Dq^2$. A relation is expected between the measure of the exchange interaction range D/T_C and the itinerant-electron character of the magnetic $3d$ electron [68]. That is, the value of D/T_C in WIE ferromagnets is larger than in localized magnetic moment systems. For example, the value of D/T_C for localized magnetic moment systems such as EuO and EuS is 0.17 and 0.16, respectively [69, 70], while D/T_C equals 2 for a typical WIE ferromagnet Ni$_3$Al [52], using the value of D_N determined by neutron scattering. For the La(Ni$_{1-x}$Al$_x$)$_{13}$ amorphous alloys, the values of D_M determined from the magnetization are respectively 140, 91 and 69 meV Å^2 for $x = 0.025$, 0.050 and 0.075, and hence the values of D/T_C become 0.63, 0.56 and 0.82, respectively. Although the value of D_M determined from the M–$T^{3/2}$ relation, especially Fe-based alloys, is different from D_N [68], the magnitude of D_M/T_C for the La(Ni$_{1-x}$Al$_x$)$_{13}$ amorphous alloys is much larger than that for localized moment systems.

Since the La(Ni$_{1-x}$Al$_x$)$_{13}$ amorphous alloys exhibit WIE ferromagnetic properties, SFs are expected to influence the temperature dependence of magnetization. In higher temperature ranges than the range governed by the spin-wave excitation, two kinds of magnetic excitation occur predominantly in WIE ferromagnets, i.e., SF and Stoner-type single-particle excitations. Both excitations give a characteristic temperature (T) dependence of magnetization (M) as mentioned in connection with (3.9) and (3.11).

The thermal variation of spontaneous magnetization $M(0, T)_{SP}$ of the La(Ni$_{0.925}$Al$_{0.075}$)$_{13}$ amorphous alloy is plotted in the form $M^2(0, T)/M^2(0, 0)$ vs. $(T/T_C)^\beta$ ($\beta = 4/3$ and 2) in Fig. 3.24. The value of $M(0, T)$ is derived from the isothermal magnetization curves. Apparently, the plots with $\beta = 4/3$ exhibit a linear relation in a wide temperature range except for low temperatures, whereas the plots with $\beta = 2$ do so only in a limited low temperature range. On the other hand, the thermal variation of $M^2(0, T)$ is quadratic for the La(Ni$_{0.975}$Al$_{0.025}$)$_{13}$ amorphous alloy, as can be seen from the inset in Fig. 3.24. The coefficient of T^2 is a measure of the contribution from thermal SFs in addition to that from Stoner-type single-particle excitations. The single-particle excitations provide a smaller contribution than the SFs to the quadratic thermal variation of $M^2(0, T)$, because they result from a thermal smearing of the Fermi surface [2]. As can be seen from (3.11), the coefficient of T^2 is obtained theoretically as $T_C^{-2}/2$ from the single-particle excitation

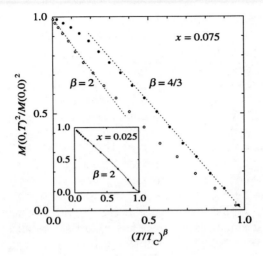

Fig. 3.24. Thermal variation of the spontaneous magnetization $M(0,T)_{\text{SP}}$ of the La(Ni$_{0.925}$Al$_{0.075}$)$_{13}$ amorphous alloy in the form $M^2(0,T)/M^2(0,0)$ vs. $(T/T_{\text{C}})^\beta$ with $\beta = 4/3$ and 2 [61]

model [64]. On the other hand, the coefficient of T^2 is enhanced by SFs via a paramagnon-like mode of fluctuations [43,71]. In the WIE ferromagnet limit, the T^2 relation due to the paramagnon-like mode of fluctuations is obtained from (3.9) as [43]

$$\frac{M^2(0,T)}{M^2(0,0)} = \frac{1}{2T_{\text{C}}^2}\left[T_{\text{C}}^2 - \left(\frac{9\pi^2}{16}\gamma + \frac{243\pi^2\psi}{8}\gamma^2\right)T^2\right] , \qquad (3.12)$$

with

$$\gamma = 0.297\frac{T_0^{1/3}T_{\text{A}}}{T_{\text{C}}} ,$$

where ψ is the Stoner enhancement factor. T_{A} and T_0 are the parameters determining the spectral widths of Im$\chi^{+-}(q,\omega)$ expressed in the double Lorentzian form

$$\text{Im}\chi^{+-}(q,\omega) = \chi_q\frac{\omega\Gamma_q}{\omega^2 + \Gamma_q^2} , \qquad (3.13)$$

with

$$\chi_q = \frac{1}{\chi^{-1} + Dq^2} , \qquad D = \frac{2T_A}{N_0q_{\text{B}}^2} ,$$

and

$$\Gamma_q = \frac{q}{\chi_q}\frac{2\pi D}{q_{\text{B}}^3}T_0 ,$$

where q_B is the boundary wave number and χ^{-1} is the inverse magnetic susceptibility. The coefficient S of the $M^2(0,T)$–ST^2 plot and the value of $T_C^{-2}/2$ are listed in Table 3.8. The estimated value of $T_C^{-2}/2$ for the La(Ni$_{0.925}$Al$_{0.075}$)$_{13}$ amorphous alloy is 0.64×10^{-4}, while the observed value is 1.86×10^{-4}, about 3 times larger. A similar comparison has been made for the La(Ni$_{0.975}$Al$_{0.025}$)$_{13}$ amorphous alloy and the ratio of the two values is about 1.5. Accordingly, an enhancement of S exists due to SFs, becoming smaller with increasing Ni concentration. Furthermore, it has been pointed out that the temperature range for the $M^2(0,T)$–T^2 dependence is made wider by contributions from both longitudinal and transverse SFs, compared to the contribution from longitudinal SFs alone. The Curie temperature T_C of the La(Ni$_{0.975}$Al$_{0.025}$)$_{13}$ amorphous alloy is 220 K, about 130 K higher than the T_C for the La(Ni$_{0.925}$Al$_{0.075}$)$_{13}$ amorphous alloy. In the former alloy, the ferromagnetic properties are further stabilized, and hence the exciting energies of both longitudinal and transverse SFs grow closer. Consequently, the T^2 dependence of $M^2(0,T)$ over a wide temperature range for the La(Ni$_{0.975}$Al$_{0.025}$)$_{13}$ amorphous alloy is explained by contributions from both longitudinal and transverse SFs.

The validity of the M^2–$T^{4/3}$ relation is confirmed in the La(Ni$_{1-x}$Al$_x$)$_{13}$ amorphous alloys with various Ni concentrations by referring to Fig. 3.25. For these curves, the in-field magnetization data $M(H,T)$ ($H = 100$ Oe) were used for brevity. The inset shows enlarged plots of the La(Ni$_{0.90}$Al$_{0.10}$)$_{13}$ amorphous alloy at temperatures below T_C. The square M^2 of the magnetization for these alloys varies linearly with $T^{4/3}$ over a relatively wide temperature range below T_C. Although SFs are influenced by the external magnetic field [56,72], the M^2–$T^{4/3}$ relation observed in the magnetic field of 100 Oe for the La(Ni$_{1-x}$Al$_x$)$_{13}$ amorphous alloys is a convincing experimental result, because the energy related to the magnetic field is too small to make a drastic change in the SFs. Strictly speaking, only a slight deviation from the straight line around T_C would be caused by the strength of the external magnetic field.

As discussed in Sect. 3.3.3, there exists an intermediate regime between weak itinerant-electron ferromagnets and localized magnetic moment limits, and both the thermal variation of S_L^2 and the spectral width of Im$\chi^{+-}(q,w)$ gradually change as the system moves from the former limit to the latter limit [62]. In fact, the M^2–$T^{4/3}$ curve is flat even at low temperatures for

Table 3.8. The Curie temperature T_C, the coefficients S and $T_C^{-2}/2$ and the ratio $S/(T_C^{-2}/2)$ for La(Ni$_{1-x}$Al$_x$)$_{13}$ amorphous alloys with $x = 0.075$ and 0.025 [61]. The parameter S is explained in the text

x	T_C [K]	S [10^{-4}K^{-2}]	$T_C^{-2}/2$ [10^{-4}K^{-2}]	$S/(T_C^{-2}/2)$
0.075	85	1.86	0.64	2.8
0.025	220	0.15	0.10	1.6

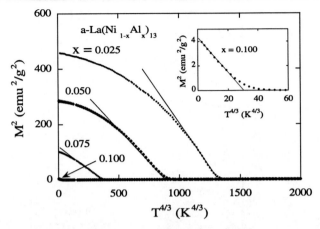

Fig. 3.25. Thermal variation of the spontaneous magnetization $M(0,T)_{\mathrm{SP}}$ of La(Ni$_{1-x}$Al$_x$)$_{13}$ amorphous alloys in the form $M^2(0,T)$–$T^{4/3}$ [61]. The *inset* is the enlarged curve of La(Ni$_{0.900}$Al$_{0.100}$)$_{13}$ amorphous alloys

the La(Ni$_{0.90}$Al$_{0.10}$)$_{13}$ amorphous alloy, indicating that the M^2–$T^{4/3}$ relation holds in a wide temperature range, while its validity deteriorates in lower temperature ranges with increasing T_{C}, being preserved only near T_{C}. These results are consistent with the concentration dependence of the $M^2(0,T)$–T^2 and M^2–$T^{4/3}$ plots given in Figs. 3.24 and 3.25. All these types of behavior are well explained by the reduced magnitude of the thermal variation of S_{L}^2 and the wider width of Im$\chi^{+-}(q,w)$ in the SF features.

Arrott Plots and Spin Fluctuation Parameters

To elucidate the change in the SF features with the concentration for the La(Ni$_{1-x}$Al$_x$)$_{13}$ amorphous alloy system, the Arrott plots at 4.2 K are displayed in Fig. 3.26. All the plots exhibit a linear variation, except for the very low magnetic field regime, and hence the parameters T_{A} and T_0 can be evaluated. The resulting values of T_{A} and T_0 are listed in Table 3.9 [61]. As shown in the table, the concentration dependences of T_{A} and T_0 indicate that the spectral width of SFs becomes broader with the Ni concentration. It should be mentioned that both T_{A} and T_0 become comparable to T_{C} in the localized magnetic moment limit, while these are 10–100 times larger than T_{C} in the WIF limit. The values of T_{C}/T_0 are of the order of 10^{-2}–10^{-1}, indicating that the itinerant-electron character is evident in the La(Ni$_{1-x}$Al$_x$)$_{13}$ amorphous alloys. The concentration dependence of both parameters indicates that the SF features shift from the WIE ferromagnet limit to the intermediate regime as the Ni concentration increases, but the longitudinal fluctuations are still dominant in the elementary excitations.

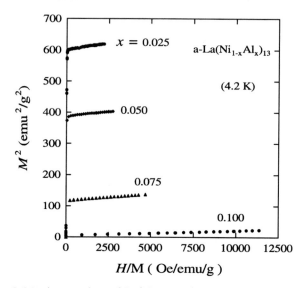

Fig. 3.26. Arrott plots of La(Ni$_{1-x}$Al$_x$)$_{13}$ amorphous alloys [41]

The $T^{4/3}$ Law in the Amorphous State

As discussed at the beginning of this section, the magnetization M near T_C obeys the M^2–$\eta T^{4/3}$ relation. The relation between the proportionality coefficient η and $M(0,0)$ and T_C is expressed theoretically by [2, 46, 53]

$$\eta = \Phi \frac{M^2(0,0)}{T_C^{4/3}} . \tag{3.14}$$

The parameter Φ is related to the $3d$ band structures at the Fermi level, but it can be considered as constant when the temperature and concentration dependences of the band structures are weak [2]. Referring to the recent theoretical discussions of SFs, this relation can be expressed in the alternative form [73]

$$\frac{M^2(0,0)}{T_C^{4/3}} = \Phi' \frac{1}{T_A T_0^{1/3}} , \tag{3.15}$$

Table 3.9. Spin fluctuation parameters T_A and T_0 for La(Ni$_{1-x}$Al$_x$)$_{13}$ amorphous alloys [61]. The parameters T_A and T_0 are explained in the text

x	T_A [10^4 K]	T_0 [10^3 K]	T_C/T_0
0.025	2.28	1.82	0.12
0.050	2.47	1.54	0.11
0.075	3.59	2.06	0.042
0.100	6.00	3.35	0.004

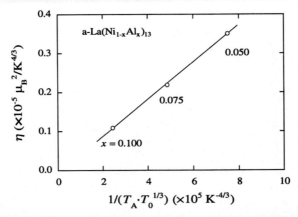

Fig. 3.27. The coefficient η against $1/T_A T_0^{1/3}$ as a function of concentration for La(Ni$_{1-x}$Al$_x$)$_{13}$ amorphous alloys [126]. All the symbols are explained in the text

where Φ' is constant under the same conditions as Φ in (3.14). Using (3.14) and (3.15), the validity of both the M^2–$T^{4/3}$ relation and the relation between the Arrott plot and the SF spectra can be confirmed. The coefficient η is plotted against $1/(T_A T_0^{1/3})$ for different concentrations x in Fig. 3.27. As can be seen from the figure, (3.14) and (3.15) are valid. This result indicates that both M near T_C and the magnetization process at low temperatures are dominated by longitudinal SFs with small q and ω. For $x = 0.025$, however, the plotted point deviates from the straight line. This deviation might be attributable to a change in SF features. To put it more concretely, the possible origins of this change are excitations of SFs with a shorter wavelength and/or the occurrence of a higher order mode–mode coupling, both being neglected in the SCR theory [2, 53].

Temperature Dependence of Paramagnetic Susceptibility

As mentioned in Sect. 3.3.2, the Curie–Weiss behavior of the inverse magnetic susceptibility χ^{-1} in WIE ferromagnets originates from the temperature dependence of SFs. The relation between χ^{-1} and S_L^2 can be derived from (3.8), (3.10) and (3.13):

$$\chi^{-1} = \frac{4N_0 I^2 S_L^2(T_C)}{3T_0} \left[S_L^2(T) - S_L^2(T_C)\right] . \tag{3.16}$$

The inverse magnetic susceptibility χ^{-1} exhibits a unique temperature dependence, as shown in Fig. 3.28. Indeed, χ^{-1} becomes convex upwards just above T_C, and subsequently undergoes an upturn. It has been pointed out that the χ^{-1}–T curve is convex upwards when the longitudinal stiffness is small, and the mean square of the amplitude of local spin density fluctuations S_L^2 exhibits a marked increase with temperature above T_C [2, 46, 53]. Therefore,

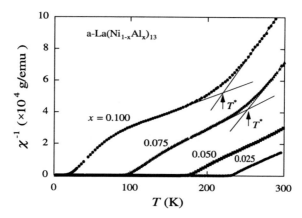

Fig. 3.28. Temperature dependence of the inverse susceptibility of La(Ni$_{1-x}$Al$_x$)$_{13}$ amorphous alloys. *Arrows* indicate the saturation temperature T^* [61]

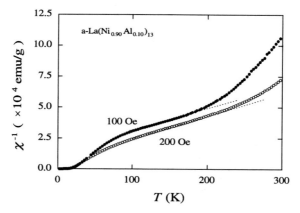

Fig. 3.29. Field effect on the temperature dependence of the inverse susceptibility for the La(Ni$_{0.90}$Al$_{0.10}$)$_{13}$ amorphous alloy [61]. *Broken lines* are guides to the eye

the longitudinal stiffness of the local spin density in La(Ni$_{1-x}$Al$_x$)$_{13}$ amorphous alloys is small and S_L^2 should rapidly increase with temperature just above T_C. Moreover, the drastic upturn in the χ^{-1}–T curves of the present amorphous alloys would be explained by the saturation of S_L^2 [62, 74].

The saturation temperature T^* indicated by the arrows in Fig. 3.28 for the La(Ni$_{1-x}$Al$_x$)$_{13}$ amorphous alloys is estimated from the intersection of the straight lines obtained from the χ^{-1}–T curve before and after changing the slope. The temperature T^* slightly increases with increasing Ni concentration. As can be seen from the figure, the SF features are influenced by the external magnetic fields. The field dependence of the χ^{-1}–T curve of the La(Ni$_{0.9}$Al$_{0.1}$)$_{13}$ amorphous alloy in fields of 100 and 200 Oe is given in Fig. 3.29 [61]. As can be seen from the figure, the χ^{-1}–T curve clearly changes

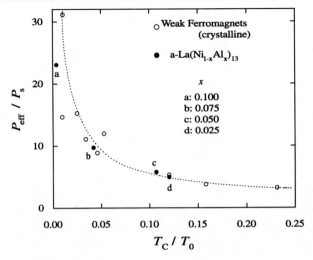

Fig. 3.30. Generalized Rhodes–Wohlfarth plots of La(Ni$_{1-x}$Al$_x$)$_{13}$ amorphous alloys [61], together with those of crystalline weak ferromagnets. The *broken line* stands for the theoretical result [73]

with the magnetic field. In other words, the curvature of the χ^{-1}–T curve between T_C and T^* is less noticeable and the upturn in the slope is reduced in a higher magnetic field. The external magnetic field tends to suppress the thermal growth of SFs [56, 72], resulting in a change in the χ^{-1}–T curve from convex upwards to linear. Needless to say, T^* is influenced by the external magnetic field, stemming from the suppression of SFs. Consequently, the change in the χ^{-1}–T curve due to the external magnetic field is clear evidence for a correlation between the characteristic properties of χ^{-1} and thermal SFs in the La(Ni$_{1-x}$Al$_x$)$_{13}$ amorphous alloys.

Generalized Rhodes–Wohlfarth Plots

The GRW plot proposed on the basis of SFs [73] is a modification of the Rhodes–Wohlfarth (RW) plot [75]. The ratio of the magnetic carrier number P_c derived from the effective moment P_{eff} to the saturation moment P_S is plotted against the Curie temperature T_C in the conventional RW plot [55]. This plot is considered to be an empirical measure of the itinerancy of magnetic materials and the SCR theory has succeeded in giving a physical meaning to the plots [2].

The GRW plot originates from advanced concepts concerning SFs, such as the correlation between the thermal variation of SFs and the magnitude of T_C/T_0 [73]. Figure 3.30 represents generalized Rhodes–Wohlfarth (GRW) plots of the La(Ni$_{1-x}$Al$_x$)$_{13}$ amorphous alloys [61], together with the theoretical broken line and data denoted by open circles for various WIE ferromagnets such as crystalline Pd–Ni alloys, ZrZn$_2$ and Ni$_3$Al [73]. The abscissa and ordinate are the ratios of the effective moment P_{eff} to the saturation moment P_S, P_{eff}/P_S, and the ratio T_C/T_0, respectively. The effective moment P_{eff} for the La$(Ni_{1-x}Al_x)_{13}$ amorphous alloys is obtained from the linear part of the χ^{-1}–T curve in a field of 100 Oe just below T^*. Although the χ^{-1}–T curve depends on the external magnetic field, the difference between the slopes of the linear part in the curves is small, as can be seen in Fig. 3.29 [61].

The ratio T_C/T_0 is one of the parameters connected with the rate of increase of S_L^2 above T_C [73], and the thermal variation of χ^{-1} is directly connected with the increase in S_L^2 [46, 53]. In Fig. 3.30, the ratio P_{eff}/P_S strongly depends on T_C/T_0 and it has been confirmed that the experimental values of various WIE ferromagnets are situated along the line in the P_{eff}/P_S–T_C/T_0 plane. The GRW plots of the La$(Ni_{1-x}Alx)_{13}$ amorphous alloys given by the solid circles are similar to those of the crystalline alloys and compounds mentioned above. Consequently, the GRW plots of the La$(Ni_{1-x}Al_x)_{13}$ amorphous alloys reveal that P_{eff}/P_S is closely connected with T_C/T_0, i.e., the variation of χ^{-1} above T_C depends on the thermal variation of S_L^2, as can be seen in (3.16).

3.4.2 La$(Co_{1-x}Al_x)_{13}$ and La$(Co_{1-x}Mn_x)_{13}$ Amorphous Alloys

Many Co-based amorphous alloys exhibit a high T_C with a large magnetic moment. According to the finite-temperature theory under the local environment effect (LEE), the ferromagnetism in the dense random packing of hard Co spheres in the amorphous state is enhanced because the main peak is near the Fermi level, so that strong ferromagnetism is realized [11]. The calculation shows that T_C in the amorphous state becomes much higher than that of Co in an fcc structure [11]. A striking enhancement of T_C has been confirmed experimentally [76]. The magnetic moment calculated on the basis of the tight-binding LMTO-recursion method for Co in the amorphous state is $1.63\mu_B$ [10], which is slightly larger than $1.58\mu_B$ for an fcc Co [77, 78] and $1.55\mu_B$ for an hcp Co [77, 79]. These values are summarized in Table 3.10.

La$(Co_{1-x}Al_x)_{13}$ [80] and La$(Co_{1-x}Mn_x)_{13}$ compounds are easily formed [81]. As mentioned in Sect. 3.2.4, the La$(Co_{1-x}Al_x)_{13}$ and La$(Co_{1-x}Mn_x)_{13}$ amorphous alloys contain icosahedral clusters. In comparison with the Ni-

Table 3.10. Calculated magnetic moment of Co in amorphous [10], fcc [77, 78], hcp [77, 79] and bcc [78] phases. R recursion, FSM fixed-spin moment, LMTO linear muffin-tin orbital

	Moment (μ_B)		Method		Reference
Amorphous	1.63		R		[10]
Fcc	1.58	1.56	LMTO	FSM	[77, 78]
Hcp	1.55		LMTO	FSM	[77, 79]
Bcc	1.68		FSM		[78]

based icosahedral-cluster amorphous alloys, the magnetic properties of Co-based alloys are relatively close to the localized magnetic moment limits. However, the substitution of Al or Mn for Co influences the $3d$ band structure. It is well known that the Co–Mn crystalline alloy system follows a specific branch of the Slater–Pauling curve, indicating that the rigid-band model is invalid. Therefore, the magnetic properties of the La(Co$_{1-x}$Mn$_x$)$_{13}$ amorphous alloys are also expected to be different from the magnetic properties predicted from the rigid-band model.

La(Co$_{1-x}$Al$_x$)$_{13}$ Amorphous Alloys

The LaCo$_{13}$ compound is formed without any additional elements. The highest Co composition limit in the amorphous state depends on the difference in radius between Co and the alloying elements. In fact, the larger the difference in radius becomes, the wider the range on the Co-rich side will be [82]. Accordingly, it is easy to prepare the LaCo$_{13}$ amorphous alloy without Al. Saturation of the magnetization curves for LaCo$_{13}$ in the amorphous and crystalline states is achieved by a relatively low magnetic field and the value in the amorphous state is larger than that in the crystalline state. The Curie temperature T_C of Co-based amorphous alloys is generally higher than the crystallization temperature in a high Co concentration range. Accordingly, T_C for the LaCo$_{13}$ amorphous alloy is scaled by the reduced magnetization curves of other amorphous Co-based alloys which have a higher crystallization temperature than T_C, such as Y–Co amorphous alloys [76].

The Curie temperatures of LaCo$_{13}$ in the amorphous and crystalline states are given in Table 3.11, together with other magnetic data. The value of T_C for the former is much higher than for the latter. The magnetic moment of the LaCo$_{13}$ amorphous alloy is larger than that of the crystalline counterpart. Structural analysis of the amorphous alloy indicates that the average environment characteristics in the nearest-neighbor range for Co–Co pairs are not so different from those in the crystalline state. The NaZn$_{13}$-type atomic structure closely resembles an fcc structure (see Fig. 3.4c). On the other hand, the interatomic distance and the coordination number for Co–La pairs are reduced by about 5% and 20%, respectively, compared with those for the crystalline counterpart (see Table 3.7). These differences of the structural characteristics would be one of the reasons for the enhancement of ferromagnetic properties.

A linear relationship in the form M versus $T^{3/2}$ for LaCo$_{13}$ in the amorphous and crystalline states is observed over a wide range of temperature and the spin-wave dispersion coefficient D is determined. As mentioned in (3.5), the spin-wave dispersion coefficient D is closely related to the band structure. Furthermore, the value of D/T_C is a measure of the exchange interaction range [68]. As shown in Table 3.11, the ratio of about 0.30 in the amorphous state is slightly smaller than the ratio of about 0.32 in the crystalline state, suggesting a shorter-range exchange interaction.

Table 3.11. The Curie temperature T_C, the magnetic moment per Co-atom M_{Co}, the spin-wave dispersion coefficient D and the spin fluctuation parameters T_A and T_0 for LaCo$_{13}$ in the crystalline and amorphous states [42]

State	Crystalline	Amorphous
T_C [K]	1290[a]	1660
M_{Co} [μ_B/Co atom]	1.56	1.66
D [meV Å^2]	418	491
T_A [K]	5655	7138
T_0 [K]	8753	11135
D/T_C	0.32	0.3

[a]K.H.J. Bushow and W.A.J.J. Velge: J. Less Common Met. **13**, 11 (1976)

The formation of well defined icosahedral clusters reduces the concentration fluctuation in contrast to other binary amorphous alloys, favorable for the discussion of SF features. On account of the enhancement of ferromagnetism in the amorphous state, SF features are expected to differ from those in the crystalline state. The spectral width of $Im\chi^{+-}(q,\omega)$ of SFs in the LaCo$_{13}$ amorphous alloy is expected to differ from that in the crystalline counterpart. The spin fluctuation parameters T_A and T_0 as measures of the width of $Im\chi^{+-}(q,\omega)$ in (3.13) for LaCo$_{13}$ in the crystalline and amorphous states are estimated from the Arrott plots at 4.2 K, referring to the correlation between these temperatures and the magnetization process in low temperatures [75]. The values of T_A and T_0 are smaller than T_C in WIE ferromagnets, while these temperatures are the same magnitude as T_C in the localized magnetic moment system. Although the correlation between these temperatures and the Arrott plots reveals some ambiguity in the region close to the localized magnetic moment system [75], the ratios of T_C/T_0 and T_C/T_A for LaCo$_{13}$ in both states are the same order for Y–Co amorphous alloys with a high concentration of Co [60], indicating that the magnetic properties are close to localized magnetic moment systems. It is noteworthy that the ratio of T_C/T_A has almost the same value in both the crystalline and amorphous states. From these results, it is considered that the fluctuation of spin density in the real space is the same magnitude in both states, because T_A correlates with the static SF features. In contrast, the ratio T_C/T_0 in the amorphous state is larger than that in the crystalline state. The difference between the ratio of T_C/T_0 in both states means that the magnetic properties of the LaCo$_{13}$ amorphous alloy are closer to the localized system than those of the crystalline counterpart.

The concentration dependence of the saturation magnetic moment M_{Co} of La(Co$_{1-x}$Al$_x$)$_{13}$ in the amorphous [44] and crystalline [80] states is shown in Fig. 3.31. The value of M_{Co} in the amorphous state is larger than that in the crystalline state, in accordance with the description given earlier. Fur-

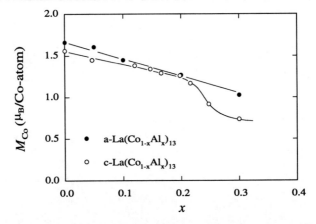

Fig. 3.31. Concentration dependence of the magnetic moment per Co atom M_{Co} of La(Co$_{1-x}$Al$_x$)$_{13}$ in the amorphous [44] and crystalline [80] states

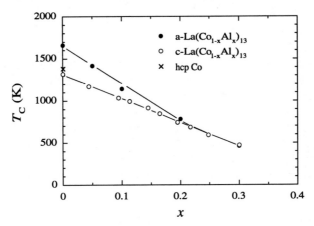

Fig. 3.32. Concentration dependence of the Curie temperature T_C of La(Co$_{1-x}$Al$_x$)$_{13}$ in the amorphous [44] and crystalline [80] states

thermore, the Curie temperature T_C of the former is higher than that of the latter, as shown in Fig. 3.32.

La(Co$_{1-x}$Mn$_x$)$_{13}$ Amorphous Alloys

The concentration dependence of the saturation magnetic moment of the La(Co$_{1-x}$Mn$_x$)$_{13}$ amorphous alloys [81] and crystalline compounds [81] is shown in Fig. 3.33, together with that of the Co$_{1-x}$Mn$_x$ crystalline alloys [83], for comparison. The magnetic moment M_{3d} in the crystalline state initially increases and then decreases rapidly with increasing x. In the amorphous state, such a broad maximum is enhanced.

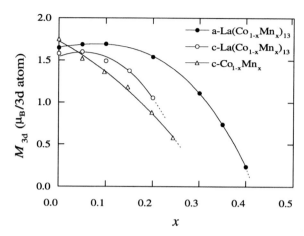

Fig. 3.33. Concentration dependence of the magnetic moment per $3d$ transition metal M_{3d} for La(Co$_{1-x}$Mn$_x$)$_{13}$ in the amorphous and crystalline states [81], together with that of Co$_{1-x}$Mn$_x$ in the crystalline state [83]

In the LaCo$_{13}$ compound, the calculated band structure [18] and experimental results such as the high value of the spin-wave dispersion coefficient and the low value of the high-field susceptibility are indications of a strong ferromagnet. La $5d$ bands are widely spread due to hybridization with Co $3d$ bands. In particular, Co $3d$ down-spin bands are hybridized with La $5d$ down-spin bands more than with up-spin bands, because Co $3d$ down-spin bands are pushed above the higher energy position by the exchange split. In a strong ferromagnetic compound, the magnetic moment is generally insensitive to the small shift in the concentration of elements on account of its filling the up-spin band. When the magnetic elements in the strong ferromagnet are substituted by others with smaller atomic number, only the electron number of the down-spin band decreases, giving rise to an increase in the magnetic moment. However, a large difference in atomic number causes not only a change in electron number, but also an ineligible local change in the potential energy around the substituted atoms. This local change in potential energy scatters plane waves of $3d$ electrons. It has been pointed out that the phase shift due to this scattering is a function of electron energy, and hence the phase shift of the scattered electron becomes just half of the incident electron phase at a certain energy, resulting in a quasi-standing wave. Due to this mechanism, electrons are trapped at a resonating energy level, which is the so-called virtual bound state (VBS) [84, 85].

When some Co is replaced by Mn in the LaCo$_{13}$ compound, the VBS forms due to the difference between the atomic numbers of Co and Mn. In the ferromagnetic state, the VBS also exhibits an exchange split because screening of the local change in potential is influenced by the difference in structure between up- and down-spin bands. At low Mn concentrations, an up-spin

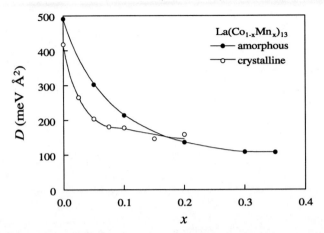

Fig. 3.34. Concentration dependence of the spin-wave dispersion coefficient D for La(Co$_{1-x}$Mn$_x$)$_{13}$ in the amorphous and crystalline states [81]

VBS of Mn is located below the Fermi level ε_F. Therefore, the decrease in electron number due to Mn substitution occurs only in the down-spin band, resulting in an increased magnetic moment. With increasing Mn concentration, the increase in local potential change and hybridization between the La $5d$ and Mn $3d$ bands push the resonating energy level of the Mn up-spin VBS above the Fermi level. Since the up-spin band has no holes in the strong ferromagnetic state, electrons in the up-spin VBS move to the down-spin band. Consequently, the magnetization begins to decrease with Mn concentration.

Turning now to Co–Mn crystalline alloys, M_{3d} decreases monotonically with x. In the Co–Mn binary system, the VBS of Mn is located at a higher energy level than or across the Fermi level because there is no hybridization between La and Mn states, and the hole number in the up-spin band increases. Then not only the magnetic moment of Mn, but also that of Co becomes small with the Mn concentration. As a result, the magnetic moment in Co–Mn drastically decreases with increasing Mn content. The spin-dependent screening mentioned above causes a variety of directional couplings between Co and Mn moments. In Co–Mn binary alloys, the magnetic moment on Co couples antiferromagnetically with that on Mn [86]. It is considered that the enhancement of ferromagnetism in the La(Co$_{1-x}$Mn$_x$)$_{13}$ amorphous alloys may also be correlated to the increase in the DOS at the Fermi level, which is caused by structural disorder [6]. These results for the La(Co$_{1-x}$Mn$_x$)$_{13}$ alloys imply that the local moments of Mn align ferromagnetically with the Co moments in lower Mn concentrations, whilst antiferromagnetic Co–Mn pairs appear in higher Mn concentrations.

The Curie temperature T_C of the amorphous alloys is higher than that of the crystalline counterparts and decreases with Mn concentration in analogy with the La(Co$_{1-x}$Al$_x$)$_{13}$ system [81]. The spin-wave dispersion coefficient

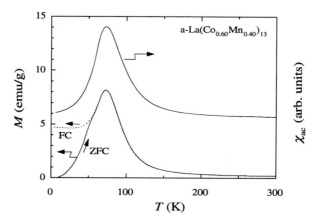

Fig. 3.35. Magnetic cooling effect in a dc magnetic field of 100 Oe and the temperature dependence of ac susceptibility for the La(Co$_{0.60}$Mn$_{0.40}$)$_{13}$ amorphous alloy [81]

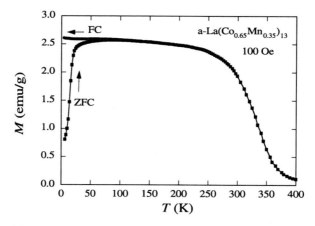

Fig. 3.36. Magnetic cooling effect in a dc magnetic field of 100 Oe for the La(Co$_{0.65}$Mn$_{0.35}$)$_{13}$ amorphous alloy [81]

D obtained from the slope of the thermomagnetization curves at low temperatures is shown in Fig. 3.34. The spin-wave dispersion coefficient D of the amorphous alloys is larger than that of the crystalline counterparts, and drastically decreases with x. However, the decrease is not so significant above about $x = 0.20$. The decrease in D implies that the spin-wave with a long wavelength is suppressed by the Mn moment, accompanied by instability of the ferromagnetic phase.

Figure 3.35 shows the magnetic field cooling effect in a dc magnetic field of 100 Oe and the temperature dependence of ac susceptibility for the amorphous alloy with $x = 0.40$. A characteristic hysteresis is observed between the zero-field-cooled (ZFC) and field-cooled (FC) states in the thermomag-

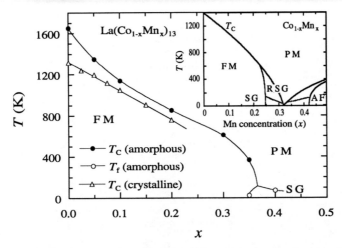

Fig. 3.37. Magnetic phase diagram of La(Co$_{1-x}$Mn$_x$)$_{13}$ in the amorphous and crystalline states [81]. The *inset* is the magnetic phase diagram of Co–Mn in the crystalline state [86]

netization curves and the cusp of ac susceptibility around 72 K. Therefore, this amorphous alloy exhibits spin-glass behavior below T_g. The appearance of the spin-glass phase is caused by frustration between ferromagnetic and antiferromagnetic interactions. It is considered that the antiferromagnetic Co–Mn pairs decrease with decreasing x, and hence the amorphous alloy with $x = 0.35$ indicates a re-entrant spin-glass (RSG) behavior. As can be seen from Fig. 3.36, the La(Co$_{0.65}$Mn$_{0.35}$)$_{13}$ amorphous alloy exhibits a magnetic field-cooling effect. The magnetic state changes from paramagnetic to ferromagnetic, and finally to the spin-glass state with decreasing temperature.

Figure 3.37 shows the magnetic phase diagram of La(Co$_{1-x}$Mn$_x$)$_{13}$ in the amorphous state, together with that for the crystalline state [81]. The inset is the magnetic phase diagram of the Co–Mn crystalline alloy system [86]. With increasing x, the ferromagnetic phase (FM) of the La(Co$_{1-x}$Mn$_x$)$_{13}$ amorphous alloy system disappears around $x = 0.35$ and the spin-glass phase (SG) appears. On the other hand, the ferromagnetic phase exists below $x = 0.25$ in Co$_{1-x}$Mn$_x$ crystalline alloys, and there is a region of frustration between ferromagnetic and antiferromagnetic interactions for $0.25 \leq x \leq 0.42$ [86]. The magnetic frustration region for the La(Co$_{1-x}$Mn$_x$)$_{13}$ amorphous alloy occurs in a higher Mn concentration range than that for Co$_{1-x}$Mn$_x$ crystalline alloys. It should thus be emphasized that the ferromagnetism of La(Co$_{1-x}$Mn$_x$)$_{13}$ in the amorphous state is more stable than that of La(Co$_{1-x}$Mn$_x$)$_{13}$ and Co$_{1-x}$Mn$_x$ in the crystalline state.

The relation between the spin-wave dispersion coefficient D and the Curie temperature T_C is given for four kinds of La(Co$_{1-x}$Mn$_x$)$_{13}$ in the crystalline and amorphous states in Fig. 3.38 [81]. As seen in (3.5), the value of D is closely correlated to the band structure. With increasing content of the ad-

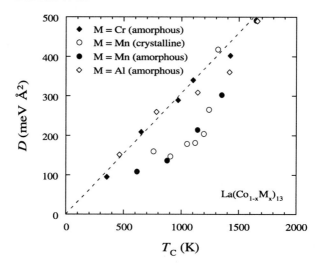

Fig. 3.38. Spin-wave dispersion coefficient D against Curie temperature T_C for La(Co$_{1-x}$Mn$_x$)$_{13}$ in the amorphous and crystalline states [81], together with that of the La(Co$_{1-x}$Al$_x$)$_{13}$ and La(Co$_{1-x}$Cr$_x$)$_{13}$ amorphous alloys [81]

ditional element, in accordance with the linear decrease in T_C, the values for La(Co$_{1-x}$Al$_x$)$_{13}$ and La(Co$_{1-x}$Cr$_x$)$_{13}$ decrease linearly. It should be noted that Cr loses its magnetic moment in the amorphous state [81]. On the other hand, Mn carries a relatively large magnetic moment, and hence the values of D for La(Co$_{1-x}$Mn$_x$)$_{13}$ in both states decrease more significantly, deviating from the linear relationship.

3.4.3 La(Fe$_{1-x}$Al$_x$)$_{13}$ Amorphous Alloys

As mentioned in Sect. 3.2.2, it is suspected that icosahedral clusters similar to those in the crystalline alloy exist in the La(Fe$_{1-x}$Al$_x$)$_{13}$ amorphous alloys. It has been suggested on theoretical grounds that $3d$ metals with an fcc structure against volume exhibit instability in their magnetic state. Figure 3.39 shows the atomic distance dependence of the magnetic moment for $3d$ fcc metals and amorphous Fe. Fcc Co shows a transition from the ferromagnetic (FM) state with magnetic moment $M_{Co} \approx 1.7\mu_B$ to the paramagnetic (PM) state with $M_{Co} = 0$ as the volume decreases [78]. Fcc Mn shows a transition from the ferromagnetic (FM) state with large M_{Mn} to the antiferromagnetic (AF) state with small M_{Mn} [87]. It should be noted that the change in magnetic state due to the volume change in fcc ferromagnetic Fe involves three kinds of magnetic state:

- the ferromagnetic high spin (HS) state with a large M_{Fe},
- the antiferromagnetic low spin (LS) state with a small M_{Fe},
- the PM state with $M_{Fe} = 0$ [78].

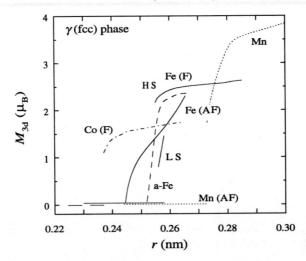

Fig. 3.39. Magnetic moment M_{3d} against interatomic distance r for Mn [87], ferromagnetic Fe [78], Co [78], antiferromagnetic Fe [89] and amorphous Fe(a-Fe) [88]. HS high spin state, LS low spin state, F ferromagnetic, AF antiferromagnetic

Note that amorphous Fe [88] also exhibits a remarkable distance dependence in analogy with γ (fcc) Fe [89]. The local symmetry of the fcc structure resembles that of the icosahedral cluster. Furthermore, the critical distance of the magnetic state for γ (fcc) Fe is close to the average distance of nearest-neighbor Fe–Fe pairs, as shown in Table 3.2. Consequently, the structural fluctuation around the average interatomic distance and the coordination number (CN) cause the variety of local magnetic states. Using the determined r and CN values, the relationship between the atomic structures and the magnetic properties of the La(Fe$_{1-x}$Al$_x$)$_{13}$ amorphous alloys are discussed in this section.

Magnetic Moment and Spin-Glass Behavior

The magnetization curves of four kinds of La(Fe$_{1-x}$Al$_x$)$_{13}$ amorphous alloy at 4.2 K are shown in Fig. 3.40. The magnetization for $x \geq 0.15$ is easily saturated in an external magnetic field less than 5 kOe, varying linearly in the high-field ranges in analogy with conventional ferromagnetic alloys. On the other hand, the curves in the concentration range $0.05 \leq x \leq 0.10$ are not easily saturated even in a field of 60 kOe. Such a phenomenon becomes more significant with increasing Fe content. This peculiar phenomenon has also been observed in many Fe-based alloys such as Zr–Fe [90], Hf–Fe [91], Sc–Fe [92] and La–Fe [94] amorphous alloys. It has been suggested that the frustrated antiferromagnetic interactions caused by the structural disorder in the amorphous state appear in the Fe-rich concentration range, resulting in the SG state [13]. Since saturation is incomplete for $x \leq 0.10$ even at 60 kOe,

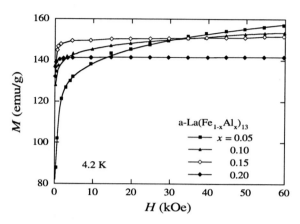

Fig. 3.40. Magnetization curves measured at 4.2 K in static fields up to 60 kOe for La(Fe$_{1-x}$Al$_x$)$_{13}$ amorphous alloys [36]

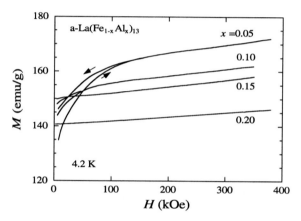

Fig. 3.41. Magnetization curves measured at 4.2 K in pulsed-magnetic fields up to 380 kOe for La(Fe$_{1-x}$Al$_x$)$_{13}$ amorphous alloys [36]. *Arrows* indicate the directions of increasing and decreasing applied magnetic fields

the magnetization curves measured up to 380 kOe using a pulse magnet for $x \leq 0.10$ seem to vary linearly with the applied field above 200 kOe, as can be seen from Fig. 3.41. The saturation magnetization M_{ST} and the high-field susceptibility χ_{hf} are obtained from the law of approach to saturation given by

$$M = M_{ST} \left(1 - \frac{a}{H} - \frac{b}{H^2} \right) + \chi_{hf} H \,, \tag{3.17}$$

where a and b are constants. The terms a/H and b/H^2 correspond to local and magnetocrystalline anisotropies, respectively. Since magnetocrystalline anisotropy is considered to be absent in amorphous alloys, M_{ST} and χ_{hf} are

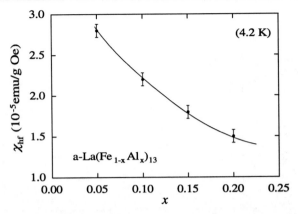

Fig. 3.42. Concentration dependence of the high-field susceptibility χ_{hf} at 4.2 K for La(Fe$_{1-x}$Al$_x$)$_{13}$ amorphous alloys [36]

determined by neglecting the b/H^2 term in (3.17). For La(Fe$_{0.95}$Al$_{0.05}$)$_{13}$ amorphous alloy, however, the $1/H$ term should be replaced by $1/H^{1/2}$ because the magnetization curve is strongly curved. This modified law of approach to saturation has been discussed by several authors [95–98].

The curves of the alloys with $x = 0.05$ and 0.10 exhibit a remarkable hysteresis associated with a spin-glass state discussed in the following section. In the case of Fe-based amorphous alloys, the exchange interaction between the nearest-neighbor Fe atoms is ferromagnetic, and the Fe atoms at farther distances couple antiferromagnetically (see the next section). Therefore, the spin-glass of the La(Fe$_{0.95}$Al$_{0.05}$)$_{13}$ amorphous alloy belongs to a cluster-type. In addition, a ferromagnetic phase is induced by applying a magnetic field (see Fig. 3.53). Under such circumstances, a power of $1/2$ is derived [95]. The concentration dependence of χ_{hf} at 4.2 K is shown in Fig. 3.42. The value of χ_{hf} for the La(Fe$_{1-x}$Al$_x$)$_{13}$ amorphous alloys increases with increasing Fe content. The concentration dependence of χ_{hf} is very similar to that of various Invar-type crystalline and amorphous alloys [99, 100].

The concentration dependence of the saturation magnetization M_{ST} for the La(Fe$_{1-x}$Al$_x$)$_{13}$ amorphous alloy system is shown in Fig. 3.43. The curve for amorphous La$_y$Fe$_{1-y}$ alloys [94] is also given in the same figure, for comparison. The value of M_S for the La(Fe$_{1-x}$Al$_x$)$_{13}$ amorphous alloys increases with increasing Fe content, approaching that of the amorphous La$_{7.5}$Fe$_{92.5}$ alloy. Note that the Fe concentration with $x = 0$ is very close to 7.5. The concentration dependence of the magnetic moment M_{Fe} of the La(Fe$_{1-x}$Al$_x$)$_{13}$ amorphous alloy system is presented in Fig. 3.44. For comparison, the data for the La(Fe$_{1-x}$Al$_x$)$_{13}$ crystalline compounds [34] and La–Fe amorphous alloys [37] are also presented in the same figure. In both the La(Fe$_{1-x}$Al$_x$)$_{13}$ and La–Fe amorphous alloy systems, the magnetic moment exhibits a similar concentration dependence, showing a broad maximum. The values of the

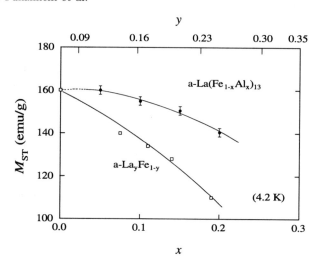

Fig. 3.43. Concentration dependence of the saturation magnetization M_{ST} of La(Fe$_{1-x}$Al$_x$)$_{13}$ amorphous alloys [36], together with that of La$_y$Fe$_{1-y}$ amorphous alloys [94]

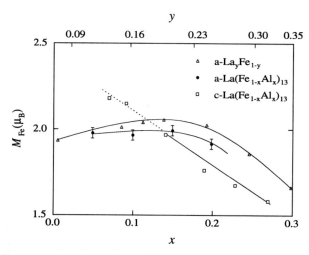

Fig. 3.44. Concentration dependence of the magnetic moment per Fe atom M_{Fe} of La(Fe$_{1-x}$Al$_x$)$_{13}$ amorphous alloys [36], together with that of the crystalline counterparts [34] and La$_y$Fe$_{1-y}$ amorphous alloys [37]

La(Fe$_{1-x}$Al$_x$)$_{13}$ amorphous alloys are lower than those of La$_y$Fe$_{1-y}$ amorphous alloys. The magnetic moment of the La(Fe$_{1-x}$Al$_x$)$_{13}$ amorphous alloys deduced from the pulsed-field measurements shows a tendency to decrease gradually towards that of the La$_{7.5}$Fe$_{92.5}$ amorphous alloy. What has to be noted is that the results obtained from the pulsed-field magnetizations for amorphous La–Fe alloys are consistent with those obtained from the Mössbauer effect measurements [37]. The concentration dependence of the magnetic moment is slightly reduced by substitution of Al for some Fe in the La(Fe$_{1-x}$Al$_x$)$_{13}$ amorphous alloys. On the other hand, the magnetic moment M_{Fe} of the La(Fe$_{1-x}$Al$_x$)$_{13}$ crystalline compounds increases linearly with increasing Fe content. In the figure, the values for the crystalline compounds in the antiferromagnetic region were determined in fields beyond the spin-flip transition [34] and are shown by the dashed line. It should be noted that the magnetic moment of the La(Fe$_{1-x}$Al$_x$)$_{13}$ amorphous alloys is smaller than that of the crystalline counterparts below about $x = 0.10$.

It is well known that the magnetic moment of Fe is significantly affected by LEEs such as the interatomic distance r of Fe–Fe pairs and the coordination number CN [101, 102]. As a consequence, the difference in the magnetic properties would originate from the difference between the above-mentioned LEEs in the amorphous and crystalline states. The correlation between the magnetic properties and the structure of Fe-rich La–Fe amorphous alloys has been investigated using small-angle and large-angle X-ray scattering [103]. According to the results, the interatomic distance of Fe–Fe atoms is slightly shortened and the mean Fe–Fe coordination number is increased with increasing Fe content. Under such circumstances, frustrated antiferromagnetic exchange interactions are considered to develop with increasing Fe content, resulting in RSG behavior at low temperatures. It has been reported that the ferromagnetic interaction is stable when the Fe–Fe coordination number in the nearest-neighbor shell of the pair distribution function is smaller than 6, while antiferromagnetic interactions prevail above this value [110].

For La(Fe$_{1-x}$Al$_x$)$_{13}$ crystalline compounds, the Fe atom sites are occupied by two different site atoms, Fe^I and Fe^{II}, as can be seen from Fig. 3.4. The Fe^I atoms are surrounded by an icosahedron of 12 Fe^{II} atoms and the Fe^{II} atoms by nine nearest Fe^{II} atoms and one Fe^I atom. Furthermore, the Fe^I–Fe^{II} distance is shorter by about 4% than the Fe^{II}–Fe^{II} distance and increases linearly with x from 0.243 nm for $x = 0.08$ to 0.251 nm for $x = 0.54$ [34]. It is worth noting that the atomic distance Fe^I–Fe^{II} in the La(Fe$_{1-x}$Al$_x$)$_{13}$ crystalline compounds [34] is much smaller than that in La$_y$Fe$_{1-y}$ amorphous alloys [103].

As can be seen from Fig. 3.40, the magnetization curves are not easily saturated in high magnetic fields for alloys in the concentration range $0.05 \leq x \leq 0.10$, and the high-field susceptibility χ_{hf} increases significantly with increasing Fe content. The results mentioned above suggest that the these alloys exhibit spin-glass-like behavior at low temperatures in this concentration range. The thermomagnetization curves of the La(Fe$_{1-x}$Al$_x$)$_{13}$ amorphous alloys are shown in Fig. 3.45. The solid lines represent the heating curves measured after cooling the samples from room temperature to 4.2 K in zero field. The broken lines show the cooling curves measured in a field of 100 Oe. For alloys in the concentration range $0.15 \leq x \leq 0.20$, the thermomagnetization curves are reversible between cooling and heating pro-

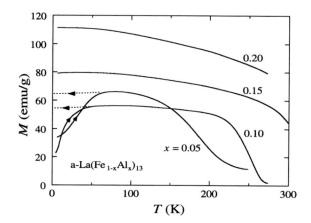

Fig. 3.45. Temperature dependence of the zero-field-cooled (ZFC) and field-cooled (FC) magnetizations in 100 Oe for La(Fe$_{1-x}$Al$_x$)$_{13}$ amorphous alloys. *Solid lines* show the zero-field cooling curves [36]

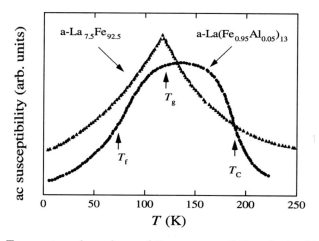

Fig. 3.46. Temperature dependence of the ac susceptibility obtained in a field of 1 Oe at 80 Hz for the La(Fe$_{0.95}$Al$_{0.05}$)$_{13}$ amorphous alloy [36], together with that for the La$_{7.5}$Fe$_{92.5}$ amorphous alloy [94]

cesses, being similar to those of conventional ferromagnets. However, alloys in the concentration range $0.05 \leq x \leq 0.10$ indicate a characteristic hysteresis between the zero-field-cooled (ZFC) and field-cooled (FC) states. The hysteresis of the field cooling effect at low temperatures becomes more significant with increasing Fe content.

It is well known that a large random magnetic anisotropy exists in amorphous alloys when they contain rare-earth metals with a non-S-state [104]. However, La is an S-state and hence the La(Fe$_{1-x}$Al$_x$)$_{13}$ amorphous alloys have no large random magnetic anisotropy. The ac susceptibility of these amorphous alloys has been investigated in order to elucidate the SG behavior. Figure 3.46 is the temperature dependence of the ac susceptibility of the La$(Fe_{0.95}Al_{0.05})_{13}$ amorphous alloy measured in an ac magnetic field of 1.0 Oe with a frequency of 80 Hz [105], together with that of the La$_{7.5}$Fe$_{92.5}$ amorphous alloy measured under the same conditions [94]. As shown in the figure, the latter exhibits a characteristic spin-glass cusp, that is to say, it undergoes a transition from a paramagnetic to a spin-glass state with decreasing temperature. On the other hand, the temperature dependence of the ac susceptibility of the former differs from that of the latter, showing a broad maximum. It has been pointed out that this broad maximum in Zr–Fe amorphous alloys divides into two peaks in superposed dc fields (see Figs. 3.51 and 3.52), which correspond to the Curie temperature T_C and the spin freezing temperature of the transverse component T_g and that of the longitudinal component T_f [106]. Accordingly, it is considered that the La$(Fe_{0.95}Al_{0.05})_{13}$ amorphous alloy system exhibits an RSG behavior.

Magnetic Phase Diagram

The magnetic phase diagram of the La$(Fe_{1-x}Al_x)_{13}$ amorphous alloy system obtained from the dc magnetization and ac susceptibility measurements is presented in Fig. 3.47, together with that in the crystalline state [34]. In the crystalline system, the magnetic phase diagram is composed of three different magnetic orders:

- in the low Fe concentration range, $0.38 \leq x \leq 0.54$, the temperature dependence curve of ac susceptibility shows a cusp, indicating a mictomagnetic-like state,
- with increasing Fe content, a ferromagnetic state appears in the concentration range $0.14 \leq x \leq 0.38$,
- in the higher Fe concentration range, $0.08 \leq x \leq 0.14$, the antiferromagnetic state is stable and the metamagnetic transition is induced by external fields of a few teslas [19].

For the amorphous system, it should be noted that the RSG behavior appears in the concentration range where antiferromagnetism develops in the crystalline system, as can be seen from Fig. 3.47. The antiferromagnetic order in the crystalline state is stabilized with increasing Fe coordination number and decreasing atomic distance in the Fe-rich concentration range, and then the Néel temperature T_N increases with increasing Fe content [107]. On the other hand, the antiferromagnetic interaction in the Fe-rich region is ascribed to high Fe–Fe coordination numbers, such as 12 for Fe^I and 10 for Fe^{II}. Furthermore, the nearest-neighbor distance of the Fe–Fe pair is very close to 0.25 nm where the antiferromagnetic interaction dominates. Antiferromagnetic long-range ordering is not easy in amorphous alloys because of the structural disorder [108]. Therefore, the SG behavior at sufficiently

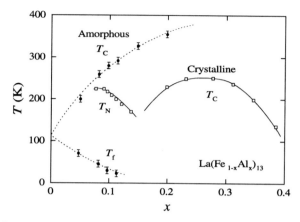

Fig. 3.47. Magnetic phase diagram of La(Fe$_{1-x}$Al$_x$)$_{13}$ in the amorphous state [36] and in the crystalline state [34]

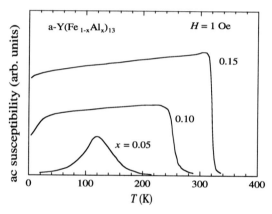

Fig. 3.48. Temperature dependence of ac susceptibility measured in a field of 1 Oe at 80 Hz for Y(Fe$_{1-x}$Al$_x$)$_{13}$ amorphous alloys [148]

low temperatures for the amorphous alloys would be due to frustrated antiferromagnetic interactions caused by structural disorder. In other words, the magnetic states in the re-entrant regime are determined by a balance between the short-range ferromagnetic interactions and the long-range antiferromagnetic interactions [6].

The RSG behavior is caused by temperature-induced enhancement of short-range interactions [6]. The Curie temperature T_C is determined from the inflection point in the high temperature ranges of the thermomagnetization curves. The re-entrant spin freezing temperature T_f is defined as the branch of these curves or the inflection point in low temperature ranges. For Zr–Fe and Y–Fe amorphous alloys, these two temperatures determined in this way correspond to the points of steepest increase and decrease in the ac sus-

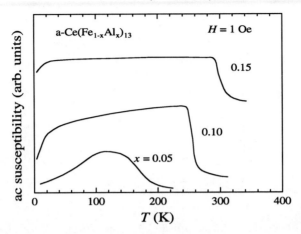

Fig. 3.49. Temperature dependence of ac susceptibility measured in a field of 1 Oe at 80 Hz for Ce(Fe$_{1-x}$Al$_x$)$_{13}$ amorphous alloys [148]

ceptibility curves measured in 1 Oe at 80 Hz [109]. Strictly speaking, T_f and T_C may be slightly higher and lower, respectively, because these temperatures are sensitive to the strength of the applied magnetic field [106, 109].

As can be seen from Fig. 3.47, T_C decreases and T_f increases with increasing Fe content, and these temperatures show a tendency to link up with the T_f value of the La$_{7.5}$Fe$_{92.5}$ amorphous alloy [93]. The point to observe is that the Curie temperature of the amorphous alloys is higher than the Curie temperature T_C and the Néel temperature T_N of the crystalline compounds, making a striking contrast to many other Fe-based alloys [13]. Figures 3.48 and 3.49 show the temperature dependence of the ac magnetic susceptibility for Y(Fe$_{1-x}$Al$_x$)$_{13}$ and Ce(Fe$_{1-x}$Al$_x$)$_{13}$ amorphous alloys, respectively. These curves are characteristic of RSGs as mentioned above, and the plateau is reduced with increasing x.

Together with additional experimental results such as the magnetic cooling effect and the differential magnetic susceptibility, the magnetic phase diagrams given in Fig. 3.50 for RE(Fe$_{1-x}$Al$_x$)$_{13}$ (RE = Y, Ce and Lu) amorphous alloys have been obtained [35]. The results for the La(Fe$_{1-x}$Al$_x$)$_{13}$ amorphous alloys are given in the same figure, for comparison. All these alloys exhibit RSG behavior below $x = 0.15$. The important point to note is that the spin freezing temperature T_f is scarcely changed by the kind of RE. This behavior is correlated with the presence of icosahedral clusters because the nearest-neighbor distance of the Fe–Fe pair hardly depends on the variety of RE (see Table 3.4). On the other hand, T_C exhibits a systematic tendency, viz., the larger the atomic size of the RE, the higher the value of T_C. The increase in the spin freezing temperature with decreasing x is related to the increase in CN of Fe. For Fe-based alloys, both M_{Fe} and T_C are strongly

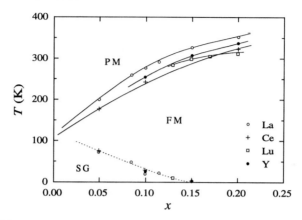

Fig. 3.50. Magnetic phase diagrams of RE(Fe$_{1-x}$Al$_x$)$_{13}$ (RE = La, Ce, Lu and Y) alloy systems [148]. P paramagnetic state, F ferromagnetic state, SG spin-glass state

affected by the local environment of the Fe atoms characterized by the CN and r values of Fe–Fe nearest-neighbor pairs.

According to the expanded finite-temperature theory taking into account the structural disorder, M_{Fe} for the central Fe atom depends on the number n of pairs defined by the central atom and the nearest-neighbor atoms located at distances shorter than the average nearest-neighbor distance [11,110]. That is, the Fe atom has a large moment of $2.5\mu_B$ for $n = 0$ and ferromagnetically interacts with moments on the neighboring atoms. The Fe atom has a small moment of $1.2\mu_B$ for $n = 6$ and interacts antiferromagnetically with moments on the neighboring atoms. The central Fe atom with $n = 12$ has no local moment [110]. Thus, the antiferromagnetic interactions in the Fe-rich region might be ascribed to the large Fe–Fe CN values of the icosahedral clusters. Furthermore, according to the relation between the magnetic state and the atomic distance for Fe [111], the short distance of the nearest-neighbor Fe–Fe pair given in Table 3.2 also introduces antiferromagnetic interactions. However, long-range antiferromagnetic order is not so feasible in amorphous alloys [108]. As a result, the spin-glass state due to frustrated antiferromagnetic interactions is expected to exist at sufficiently low temperatures [105] in the concentration range where the antiferromagnetic order appears in the crystalline state. The increase in T_f with decreasing x is related to the increase in CN from 8.4 for $x = 0.20$ to 9.5 for $x = 0.05$, as can be seen from Fig. 3.8.

The magnetic properties of Fe-based amorphous alloys which exhibit spin-glass behavior are very sensitive to the applied magnetic field and hydrostatic pressure [112]. The temperature dependence of the differential magnetic susceptibility dM/dH of the La(Fe$_{0.95}$Al$_{0.05}$)$_{13}$ amorphous alloy obtained at various dc fields is given in Fig. 3.51. The dM/dH curve measured in a field

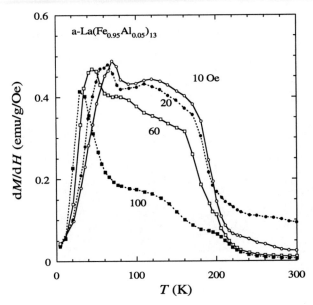

Fig. 3.51. Temperature dependence of the differential magnetic susceptibility dM/dH measured at 10, 20, 60 and 100 Oe for the La(Fe$_{0.95}$Al$_{0.05}$)$_{13}$ amorphous alloy [148]

of 10 Oe shows distinct two peaks. This means that the spin-glass transition proceeds through two steps, viz., the higher temperature T_g of the freezing temperature of the transverse component of spins and the lower temperature T_f of the longitudinal spin component [109]. Increasing the external magnetic field, both T_f and T_g shift to lower temperatures and the former peak becomes sharp, whereas the latter peak becomes uncertain. With further increase in the magnetic field, as shown in Fig. 3.52, T_g disappears and a third broad peak appears around 200 K. This is identified as T_C because it corresponds to the inflection point of the thermomagnetization curves measured in the same magnetic field. As can be seen from Fig. 3.52, this third peak shifts slightly to higher temperature with increasing applied magnetic field.

From these procedures, the relation between the freezing temperatures T_f and T_g, and the Curie temperature T_C versus the external magnetic field is obtained as presented in Fig. 3.53 for La(Fe$_{0.95}$Al$_{0.05}$)$_{13}$ and La(Fe$_{0.90}$Al$_{0.10}$)$_{13}$ amorphous alloys. The temperature dependence of dM/dH for the La(Fe$_{0.80}$Al$_{0.20}$)$_{13}$ amorphous alloy which has no spin-glass behavior is also given in Fig. 3.54. Increasing the strength of the applied magnetic field, the peak corresponding to T_C increases slightly, as shown by the dashed line. From these results, it should be emphasized that T_g and T_f are drastically decreased by relatively low magnetic fields, whereas T_C is slightly increased by applying a magnetic field.

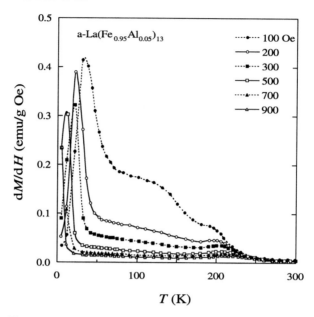

Fig. 3.52. Temperature dependence of the differential magnetic susceptibility dM/dH measured at higher magnetic fields for the La(Fe$_{0.95}$Al$_{0.05}$)$_{13}$ amorphous alloy [148]

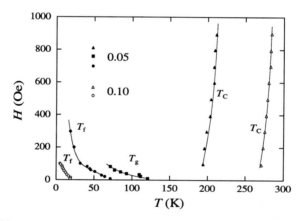

Fig. 3.53. Magnetic phase diagram of two La(Fe$_{1-x}$Al$_x$)$_{13}$ ($x = 0.05$ and 0.10) amorphous alloys in magnetic fields [36, 148]

Spin-Wave Dispersion Coefficient

As will be described in Sect. 3.5.3, the La(Fe$_{1-x}$Al$_x$)$_{13}$ amorphous alloys show a marked magnetovolume effect, resulting in the Invar characteristics. The magnetization of the La(Fe$_{1-x}$Al$_x$)$_{13}$ amorphous alloys obeys the $T^{3/2}$ temperature dependence at low temperatures only in the concentration range

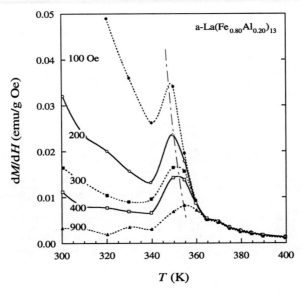

Fig. 3.54. Temperature dependence of the differential magnetic susceptibility dM/dH for the La(Fe$_{0.80}$Al$_{0.20}$)$_{13}$ amorphous alloy. The *dash-dotted line* shows the shift in the Curie temperature [148]

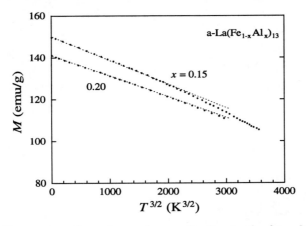

Fig. 3.55. Temperature dependence of magnetization in the form M–$T^{3/2}$ measured in 10 kOe for La(Fe$_{1-x}$Al$_x$)$_{13}$ ($x = 0.15$ and 0.20) amorphous alloys [36]. The *chain lines* are guides to the eye

$0.15 \leq x \leq 0.20$, as shown in Fig. 3.55. The straight chain line is a guide to the eye. On the other hand, such a linear dependence is not held below $x = 0.10$ due to spin freezing, as can be seen from Fig. 3.56. The spin-wave dispersion coefficient D determined from (3.6) for the La(Fe$_{1-x}$Al$_x$)$_{13}$ amorphous alloys ($0.15 \leq x \leq 0.20$) is plotted in Fig. 3.57 [36], and compared with

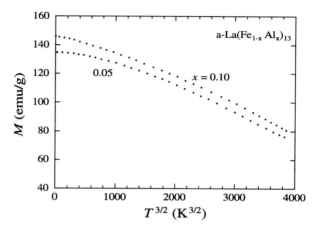

Fig. 3.56. Temperature dependence of magnetization in the form M–$T^{3/2}$ measured in 10 kOe for La$(\mathrm{Fe}_{1-x}\mathrm{Al}_x)_{13}$ ($x = 0.05$ and 0.10) amorphous alloys [36]

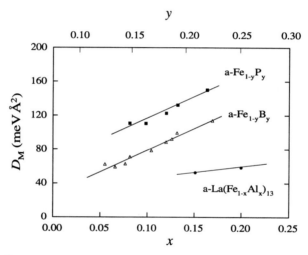

Fig. 3.57. Concentration dependence of the spin-wave stiffness constant D_M at 4.2 K for La$(\mathrm{Fe}_{1-x}\mathrm{Al}_x)_{13}$ amorphous alloys [36], together with that for Fe–B and Fe–P amorphous alloys [113]

the data for Fe–B and Fe–P amorphous alloys [113]. It should be noted that the value of D for the La$(\mathrm{Fe}_{1-x}\mathrm{Al}_x)_{13}$ amorphous alloys is extremely small, implying instability of ferromagnetism.

A conventional way to determine D is to examine the thermoagnetization curve, as mentioned in (3.6). The value of D can also be determined using the Brillouin scattering technique. For the Damon–Eshbach (DE) mode and the bulk spin waves from a series of standing spin waves (SSWs), frequencies are given as follows. For the DE mode,

$$2\pi\Delta\nu_S = \tag{3.18}$$
$$\gamma\left\{H + 2\pi M_S[1 - \exp(-Q_SL)]\right\}^{1/2}\left\{H + 2\pi M_S[1 + \exp(-Q_SL)]\right\}^{1/2},$$

and for the nth SSW mode,

$$2\pi\Delta\nu_B(n) = \tag{3.19}$$
$$\gamma\left\{H + D[Q_S^2 + (n\pi/L)^2]\right\}^{1/2}\left\{H + D[Q_S^2 + (n\pi/L)^2 + 4\pi M_B]\right\}^{1/2},$$

with

$$Q_S = 4\pi\sin\theta/\lambda_0 , \quad \omega_R = 2\pi\Delta\nu ,$$

where γ is the gyromagnetic ratio, g is the g-factor, L is the film thickness, M_S and M_B are the surface and bulk magnetizations, respectively, λ_0 is the vacuum wavelength of the laser light, and q is the scattering angle measured from the surface normal. ω_R is the resonance frequency and $\Delta\nu$ is the Brillouin shift. The values of D determined from the thermomagnetization curve and the Brillouin scattering are hereafter denoted by D_M and D_{SW}, respectively. The value of D_M for $x = 0.20$, i.e., a-La(Fe$_{0.80}$Al$_{0.20}$)$_{13}$, is 60 meV Å^2 in Fig. 3.57. At first glance, D_M exhibits a good agreement with D_{SW} of 63 ± 0 meV Å^2 at room temperature in Table 3.12. However, according to the spin-wave theory, D generally depends on T in the following way:

$$D(T) = D(0)\left[1 - \varphi\left(\frac{T}{T_C}\right)^{5/2}\right] . \tag{3.20}$$

The coefficient φ has been determined experimentally to be of the order of 0.5–0.8 [117,118]. Taking into account the temperature dependence, the value $D_{SW}(0)$ at 0 K is estimated to be 80–110 meV Å^2 and the ratio $D_{SW}(0)/D_M$ to be 1.4–1.8. Since the La(Fe$_{1-x}$Al$_x$)$_{13}$ amorphous alloys exhibit Invar characteristics (see Fig. 3.64), the estimated ratio is close to the reported values of

Table 3.12. Several kinds of magnetic constants of the La(Fe$_{0.80}$Al$_{0.20}$)$_{13}$ amorphous alloy [114], the Fe$_{80}$B$_{20}$ amorphous alloy [115] and an α (bcc) Fe [116]

	a-La(Fe$_{0.80}$Al$_{0.20}$)$_{13}$	a-Fe$_{80}$B$_{20}$	c-Fe($\equiv \alpha$-Fe)
T/T_C	0.76	0.46	0.29
g	2.10 ± 0.04	2.10	2.1
$\gamma/2\pi$ [GHz/kOe]	2.94 ± 0.06	2.94	2.94
$4\pi M_S$ [kG]	11.0 ± 0.3	–	21.5
$4\pi M_B$ [kG]	10.8 ± 0.3	15.0	21.5
D [meV Å]	63 ± 10	170	284
A [10^{-6} erg/cm]	0.22 ± 0.03	0.84	2.00

1.6–2.3 for various Invar-type Fe-based amorphous ferromagnets, in contrast to the values close to unity for non-Invar-type Fe-based amorphous ferromagnets [113]. The origin of such a large departure of the ratio from unity in the Invar-type materials is not yet fully understood. As a possible origin, an additional low-lying magnetic excitation with quadratic dispersion has been proposed [119, 120]. No additional magnetic excitations are confirmed in the spin-wave spectra. When we discuss the magnetic properties of ferromagnetic materials, e.g., the domain-wall structure, the exchange stiffness A defined by the following expression is more useful than D itself:

$$A = 4\pi M \frac{D}{8\pi} .$$
(3.21)

We give A at $290\,\mathrm{K}$ in Table 3.12, in which M_B and M_S are the bulk and surface magnetizations and g is the g-factor. Again, A for the alloy with $x = 0.20$ is much smaller than that for $\mathrm{Fe_{80}B_{20}}$ amorphous (a-$\mathrm{Fe_{80}B_{20}}$) and pure crystalline α (bcc) Fe (c-Fe) films.

3.5 Magnetovolume Effects in $\mathrm{La(TM_{1-x}Al_x)_{13}}$ Amorphous Alloys

Experimentally, information on dynamical properties of SFs is obtained by spectroscopy such as NMR and neutron scattering. It should be stressed that conventional thermal expansion measurements also give useful information on SFs. In this section, the magnitude of thermal variations of SF amplitudes in the ferromagnetic state is discussed using the thermal expansion curves in the ferromagnetic temperature ranges. The concentration and temperature dependences of the spontaneous volume magnetostriction are also discussed. Furthermore, the thermal expansion anomaly in paramagnetic temperature regions is discussed in terms of the saturation of SF amplitudes.

3.5.1 Thermal Variation of Spin Fluctuation Amplitudes

The volume of ferromagnets is sensitive to SF amplitudes and theories of SFs give a reasonable description of the thermal expansion anomaly including the Invar effect observed in some ferromagnetic alloys and intermetallic compounds [121–123]. The free energy coupled with the magnetovolume effect F_mv is given by [123]

$$F_\mathrm{mv} = -\Omega \sum_q K_q S_q^2 ,$$
(3.22)

where Ω is the volume strain, and K_q and S_q are the magnetovolume coupling constant and the spin density, respectively, for the SF component with

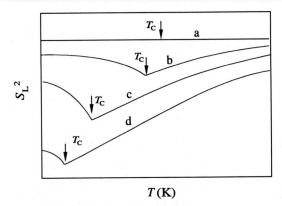

T (K)

Fig. 3.58. Temperature variations in the mean-square local amplitude of spin fluctuations S_L^2 [2, 124]. See text for details

wavenumber q. By neglecting the q dependence of K_q, the spontaneous volume magnetostriction ω_S is derived from the thermodynamical relation between volume strain and free energy as

$$\omega_s = \kappa C \left[S_L^2(0) - S_L^2(T_C) \right] \tag{3.23}$$
$$= \Omega + \kappa P,$$

where κ is the compressibility, C the magnetoelastic coupling constant and P the pressure. Consequently, the thermal expansion is an important quantity for investigating the temperature dependence of $S_L^2(T)$. Various temperature dependences of $S_L^2(T)$ are given schematically in Fig. 3.58. Since $S_L^2(T)$ is present in the paramagnetic temperature region, the magnetovolume effect also appears even above T_C [123]. This behavior cannot be explained by the Stoner theory, because the single-particle excitations completely vanish at T_C [3,46,53]. The amplitude of $S_L^2(T)$ in the paramagnetic temperature region increases with temperature, in contrast to the variation below T_C [46,47]. As a result, the contribution of $S_L^2(T)$ can be observed as a positive additional term to the thermal expansion coefficient. Note that $S_L^2(T)$ is invariable against temperature in the localized magnetic moment model as given by the line (a). Furthermore, the curves tend to be saturated with a positive curvature at high temperatures as shown by the curves (b)–(d) in Fig. 3.58 [124].

3.5.2 Thermal Expansion
Below and Above the Curie Temperature
for La(Ni$_{1-x}$Al$_x$)$_{13}$ Amorphous Alloys

The thermal expansion curves from 4.2 to 600 K for the La(Ni$_{1-x}$Al$_x$)$_{13}$ amorphous alloys are presented in Fig. 3.59. All the curves increase with temperature over the entire temperature range. At low temperatures, the thermal expansion coefficient becomes smaller with increasing Ni concentration in both

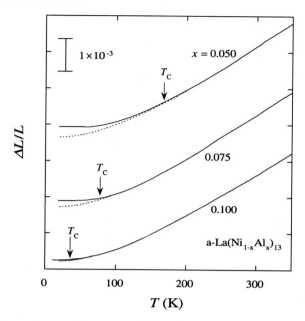

Fig. 3.59. Thermal expansion curves of for La(Ni$_{1-x}$Al$_x$)$_{13}$ amorphous alloys. The *dotted line* stands for a hypothetical paramagnetic curve contributed from electron and phonon terms [126]

systems, indicating the existence of a small magnetic contribution. In order to estimate the contribution from SFs, the contributions from phonons and electrons should be subtracted from the observed thermal expansion curves. That is, the thermal expansion coefficient α is given by

$$\alpha = \alpha_p + \alpha_e + \alpha_m , \tag{3.24}$$

where α_p, α_e and α_m are the phonon, electronic and magnetic terms, respectively. The phonon term α_p is calculated from the following Grüneisen relation between α_p and the specific heat C_V [125]:

$$\alpha_p = \Gamma \kappa C_V , \tag{3.25}$$

with

$$C_V = 9 N_A \kappa_B \left(\frac{T}{\theta_D}\right)^3 \int_0^{\theta_D/T} \frac{x^4}{(e^x - 1)^2} \mathrm{d}x ,$$

where Γ, κ, θ_D, N_A and k_B are the Grüneisen parameter, the compressibility, the Debye temperature, the number of atoms and the Boltzmann constant, respectively. After determining both α_p and α_e, the magnetic contribution α_m can be evaluated from (3.24), when the thermal expansion is affected by SFs. From (3.24), the thermal expansion curve of a hypothetical paramagnetic

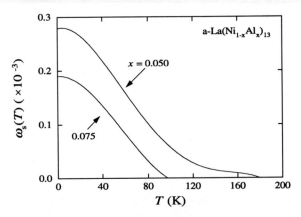

Fig. 3.60. Temperature dependence of the spontaneous volume magnetostriction $\omega_S(T)$ for La(Ni$_{1-x}$Al$_x$)$_{13}$ amorphous alloys [126]

state is fitted to the observed curves as displayed by the dotted lines in Fig. 3.59. The Debye temperature of the La(Ni$_{1-x}$Al$_x$)$_{13}$ amorphous alloy system was determined to be 280 K from the low-temperature specific heat data using the expression

$$\beta(T) = \frac{12}{5}\pi^4 N_A \kappa_B \left(\frac{1}{\theta_D(T)}\right)^3 , \tag{3.26}$$

where $\beta(T)$ is the coefficient of the phonon term. All the curves of the La(Ni$_{1-x}$Al$_x$)$_{13}$ amorphous alloys deviate from the calculated curve below T_C. The amplitude of SFs in the ferromagnetic state decreases with the temperature, and the decrement in amplitude gives a spontaneous volume magnetostriction $\omega_S(T)$ [123]. Therefore, the deviation from calculated values is closely related to ω_S.

The temperature dependence of $\omega_S(T)$ for the La(Ni$_{1-x}$Al$_x$)$_{13}$ amorphous alloys is shown in Fig. 3.60. The value of ω_S decreases with temperature and vanishes at slightly higher temperature than T_C. Accordingly, the SF amplitudes decrease in the temperature range below T_C. The value of ω_S at 0 K, $\omega_S(0)$, becomes larger with increasing Ni concentration. The value of $\omega_S(0)$ depends on the longitudinal stiffness of SFs [2, 123], because the small longitudinal stiffness of SFs leads to a large value given by

$$\Delta S_L^2 = S_L^2(0) - S_L^2(T_C) . \tag{3.27}$$

When the longitudinal stiffness is the same magnitude in the specimens, $\omega_S(0)$ is nearly proportional to the spontaneous magnetic moment M_{Ni}. Hence, the longitudinal stiffness is scarcely changed by the Ni concentration. On the other hand, the concentration dependence of $\omega_S(0)$ for the La(Ni$_{1-x}$Al$_x$)$_{13}$ amorphous alloys differs slightly from that of M_{Ni}, implying the concentration dependence of the longitudinal stiffness.

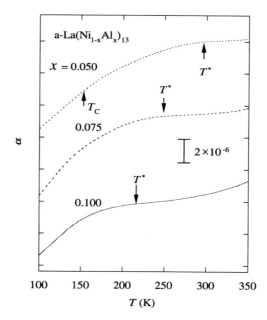

Fig. 3.61. Temperature dependence of the thermal expansion coefficient α of La(Ni$_{1-x}$Al$_x$)$_{13}$ amorphous alloys [126]

According to the SCR theory, the amplitude of SFs increases with temperature above T_C [2, 46, 53]. Therefore, the contribution of SFs to the thermal expansion coefficient is positive in the paramagnetic temperature region [123]. Because the inverse magnetic susceptibility is expected to be proportional to the amplitude of SFs, the thermal expansion due to the thermal variation of SFs is considered to be proportional to the inverse magnetic susceptibility χ^{-1} [2]. What should be noted is the peculiar behavior of the inverse magnetic susceptibility for the La(Ni$_{1-x}$Al$_x$)$_{13}$ ($x = 0.05$, 0.075 and 0.10) amorphous alloys, indicating the saturation of thermal growth of SFs. The evidence for the saturation of thermal growth of SFs should be detected by careful analyses of the thermal expansion curve in the paramagnetic temperature region.

Figure 3.61 shows the temperature dependence of the linear thermal expansion coefficient α of the La(Ni$_{1-x}$Al$_x$)$_{13}$ amorphous alloys [126]. The increase in α in the vicinity of 100 K is due to the contribution from phonons α_p and the straight variation around 350 K indicates that α_p becomes flat against temperature and the contribution from electrons α_e becomes evident for the alloy with $x = 0.100$. The thermal variation of α_p and α_e has already been discussed in connection with (3.24). The variations of α_p and α_e are illustrated schematically in Fig. 3.62. The contribution from SFs to α_m is also indicated in the same figure. The variation of α_m corresponds to the derivative of the curve for the weak itinerant-electron ferromagnet

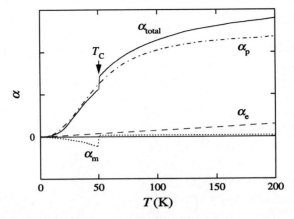

Fig. 3.62. Schematic temperature variations of the thermal expansion coefficient of the electronic term α_e, phonon term α_p and magnetic term α_m [125]

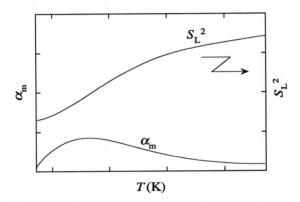

Fig. 3.63. The relation between the thermal expansion coefficient of the magnetic term α_m and the mean-square local amplitude of spin fluctuations S_L^2 as a function of temperature [41]

limit in Fig. 3.58, because α_m is proportional to dS_L^2/dT [123]. When S_L^2 is proportional to temperature as discussed in the SCR theory [2, 46, 62], the value of α_m in the paramagnetic temperature regions is constant. As a result, the increment in the total linear thermal expansion coefficient α should become sluggish with increasing temperature. One should remember that saturation of SF amplitudes is expected from the characteristic change in the inverse magnetic susceptibility of the La(Ni$_{1-x}$Al$_x$)$_{13}$ amorphous alloys (see Fig. 3.28). The saturation temperature T^* indicated by the arrow was cited from Fig. 3.28. In fact, the change in the increment in α appears in the vicinity of T^* for the La(Ni$_{1-x}$Al$_x$)$_{13}$ amorphous alloys, especially for the alloy with $x = 0.100$. Therefore, this change should be related to the peculiar temperature dependence of the SF amplitudes.

Referring to the proportionality relation between the thermal expansion and the SF amplitude S_L^2, the magnetic contribution to the linear thermal expansion coefficient α_m is illustrated in Fig. 3.63 by taking the saturation of S_L^2 [74] into account. As can be seen from this figure, the coefficient α_m increases with temperature below T^* and begins to decrease as S_L^2 approaches the saturation value. Consequently, the observed variation of the thermal expansion coefficient against temperature for the $La(Ni_{1-x}Al_x)_{13}$ amorphous alloys in paramagnetic temperature ranges should be connected with the saturation of the thermal growth of SFs.

3.5.3 Large Thermal Expansion Anomaly in $La(Fe_{1-x}Al_x)_{13}$ Amorphous Alloys

It has been reported that the $La(Fe_{1-x}Al_x)_{13}$ compounds exhibit remarkable magnetovolume effects [34]. In general, alloys with a very low thermal expansion coefficient around room temperature due to the large magnetovolume effect are called Invar alloys. These alloys exhibit various peculiarities in their magnetic properties, such as a large high-field susceptibility χ_{hf}, small spin-wave dispersion coefficient D, large pressure effect on T_C, large compressibility and significant pressure dependence of magnetization [12]. It should be noted that significant magnetovolume effects are common to Fe-based amorphous alloys [12,127]. The concentration dependences of χ_{hf} and T_C are analogous to those of Fe–B and Fe–P amorphous alloys [99], as well as Fe–Ni crystalline Invar alloys [128]. Moreover, the D value of the $La(Fe_{1-x}Al_x)_{13}$ amorphous alloys is 50–60 meV $Å^2$, much smaller than that of Fe–B and Fe–P amorphous alloys [113], as shown in Fig. 3.57. These phenomena are caused by instability of ferromagnetism. Therefore, the $La(Fe_{1-x}Al_x)_{13}$ amorphous alloys are expected to exhibit significant magnetovolume effects.

Spontaneous Volume Magnetostriction

The thermal expansion curves of five $La(Fe_{1-x}Al_x)_{13}$ amorphous alloys are given in Fig. 3.64 [105]. The broken line is the paramagnetic thermal expansion curve estimated from the Grüneisen relation, assuming that the Debye temperature $\theta_D = 300$ K, very close to the value estimated from the Brillouin scattering. These curves exhibit a significant anomaly in a wide temperature range. Figure 3.65 shows the concentration dependence of the spontaneous volume magnetostriction $\omega_S(0)$ for the same amorphous alloys.

The values of $\omega_S(0)$ are three times the difference at 0 K between the solid and broken lines in Fig. 3.64. The value of $\omega_S(0)$ decreases with increasing x, especially below 0.15, comparable to those for Fe–Ni fcc crystalline Invar alloys [129]. The variation in the spontaneous volume magnetostriction $\omega_S(T)$ against the reduced temperature T/T_C for the $La(Fe_{1-x}Al_x)_{13}$ amorphous alloys is given in Fig. 3.66. The value of $\omega_S(T)$ increases with increasing Fe

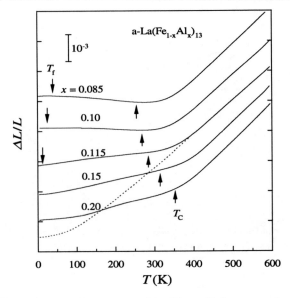

Fig. 3.64. Thermal expansion curves of La(Fe$_{1-x}$Al$_x$)$_{13}$ amorphous alloys. The Curie temperature T_C and the spin freezing temperature T_f are given by the *arrows*. The *broken line* stands for a hypothetical paramagnetic curve [105]

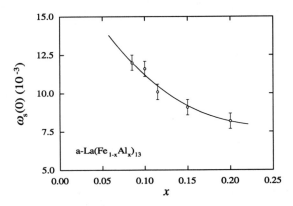

Fig. 3.65. Concentration dependence of the spontaneous volume magnetostriction at 0 K, $\omega_S(0)$, for La(Fe$_{1-x}$Al$_x$)$_{13}$ amorphous alloys [105]

content. Furthermore, it is evident that a marked thermal expansion anomaly still remains, even well above T_C. A similar behavior has also been observed for many other Fe-based amorphous alloys [12, 13].

For homogeneous ferromagnetic alloys, the Arrott plots exhibit straight lines with slope independent of temperature, while deviations give an indication that the materials are magnetically inhomogeneous [130]. Figures 3.67 and 3.68 show the Arrott plots for La(Fe$_{0.80}$Al$_{0.20}$)$_{13}$ ferromagnetic and

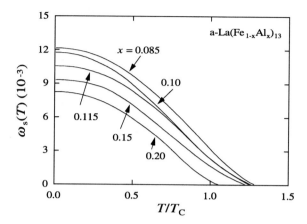

Fig. 3.66. Spontaneous volume magnetostriction of La(Fe$_{1-x}$Al$_x$)$_{13}$ amorphous alloys as a function of the reduced temperature T/T_C [105]

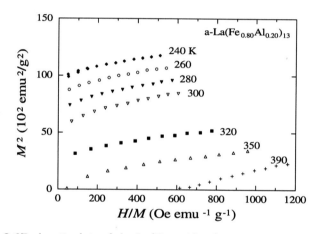

Fig. 3.67. Arrott plots of the La(Fe$_{0.80}$Al$_{0.20}$)$_{13}$ amorphous alloy [171]

La(Fe$_{0.90}$Al$_{0.10}$)$_{13}$ RSG amorphous alloys, respectively. These plots show strongly curved lines, reflecting an inhomogeneous magnetic state. The Arrott plots for inhomogeneous systems have been discussed using a variant of the Landau–Ginzburg (LG) theory taking into account cooperative SFs [131], and also using a superparamagnetic model [132]. To a certain extent, such an inhomogeneity would be reflected in the thermal expansion anomaly. However, a distinct anomalous thermal expansion above T_C has even been observed in a homogeneous Y$_2$Fe$_{17}$ crystalline compound annealed at 1373 K for 3 days [133]. In addition, a negative thermal expansion, i.e., a shrinkage, still exists even well above T_C for alloys below $x = 0.10$, as shown in Fig. 3.64.

Theoretically and experimentally, on the other hand, it has been pointed out that SFs play an important role in magnetic properties regardless of inho-

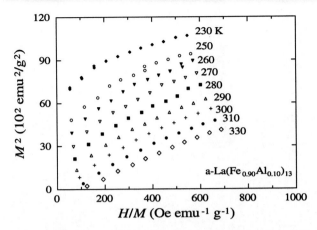

Fig. 3.68. Arrott plots of the La(Fe$_{0.90}$Al$_{0.10}$)$_{13}$ amorphous alloy [105]

mogeneity [11,60,61,110,134,135]. It has been pointed out that the anomalous thermal expansion in Fe–Ni fcc Invar alloys is mainly caused by the large temperature variation in amplitude of the Fe local moments [130]. Because the Fe-rich Invar alloys are considered to be intermediate ferromagnets between strong and very weak ferromagnetic materials, the Coulomb interaction plays an important role. Therefore, a negative thermal expansion would be brought about by a large contraction in amplitude of the local moment caused by an increase in temperature. In fact, this type of peculiar phenomenon has been presented in the calculation for Fe$_{1-x}$Ni$_x$ (fcc) alloys such as $x = 0.20$, even though this alloy is in the paramagnetic state [136].

Electronic Contribution to the Thermal Expansion Anomaly

Let us devote a little more space to discussing the peculiar thermal expansion mentioned above. The finite-temperature theory of magnetism for amorphous TMs [6,137] is based on a full-static single-site approximation with a functional integral method [138–140], and SFs in amorphous TMs with narrow bands have been effectively investigated. Magnetic properties of Fe-rich alloys are closely associated with the LEEs [11], and hence the mean-squared amplitude of the local moment $\langle m^2 \rangle$ depends predominantly on the surrounding environment [11], where $\langle \ \rangle$ denotes the thermal average. The spontaneous volume magnetostriction $\omega_S(T)$ is correlated to $\langle m^2 \rangle$. Neglecting the contribution of s–d charge transfer [122,141], it is expr essed by

$$\omega_S(T) - \omega_S(0) \approx \frac{\Gamma_e \kappa}{3V} \int_0^T C_V^e dT + \frac{1}{4} U \sum_i \left(\langle m_i^2 \rangle \big|_T - \langle m_i^2 \rangle \big|_0 \right) , \quad (3.28)$$

where κ, V, Γ_e and U are the compressibility, the bulk volume, the electronic and magnetic Grüneisen parameter and the effective Coulomb en-

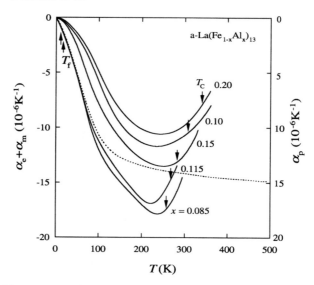

Fig. 3.69. Temperature dependence of the electronic and magnetic contributions to the thermal expansion coefficients $\alpha_e + \alpha_m$ for the La(Fe$_{1-x}$Al$_x$)$_{13}$ amorphous alloys. The *broken line* shows the phonon contribution α_p to the thermal expansion coefficient [35]

ergy, respectively. In (3.28), C_V^e denotes the electronic and magnetic specific heat [122, 141]. Peculiar behavior such as a negative thermal expansion is explained by the contribution of $\langle m^2 \rangle$ [141]. The increase in $\langle m^2 \rangle$ with decreasing temperature corresponds to the increase in $\omega_S(T)$.

The electronic and magnetic contribution to the linear thermal expansion coefficient $\alpha_e + \alpha_m$ is plotted against temperature for the La(Fe$_{1-x}$Al$_x$)$_{13}$ amorphous alloys in Fig. 3.69. The value of $\alpha_e + \alpha_m$ is obtained from the temperature dependence of $\omega_S(T)$. The phonon contribution α_p with a positive sign, which is obtained from the dotted curve in Fig. 3.64, is also shown by the dashed line in the same figure. For example, the magnitude of $\alpha_e + \alpha_m$ for the La(Fe$_{0.95}$Al$_{0.05}$)$_{13}$ amorphous alloy is larger than α_p below about 280 K. Therefore, the total linear thermal expansion coefficient is negative. In other words, the volume contraction occurs with increasing temperature up to 280 K, accompanied by a minimum in $\alpha_e + \alpha_m$ in the paramagnetic temperature range, as can be seen from Fig. 3.69. Such a peculiar temperature dependence of the value of $\alpha_e + \alpha_m$ is similar to that of Fe–Ni crystalline alloys [141].

The theoretical results for the temperature dependence of $\alpha_e + \alpha_m$ for Fe–Ni crystalline alloys also explain the experimental results for the La(Fe$_{1-x}$Al$_x$)$_{13}$ amorphous alloys, because the temperature dependence of $\alpha_e + \alpha_m$ is mainly associated with the amplitude of the local moment for the Fe site as well as Fe-rich amorphous alloys [122, 141]. This theory explains

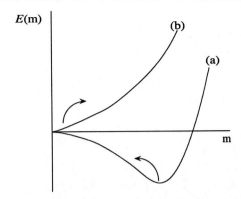

Fig. 3.70. Schematic diagram of the magnetic energy $E(m)$ against the local moment m [122]. (a) the ground state, (b) the paramagnetic state

that the thermal excitation occurs from the state with a large amplitude of moment to that with a small one, in the paramagnetic temperature region [141]. The excitation depends on the variation of the magnetic energy as a function of the local moment m [141]. As mentioned above, the amplitude of m depends on the r and CN values of $3d$ metals in itinerant-electron systems, so that features of the magnetic energy curve in amorphous alloys change depending on the number of Fe atoms, which is contracted and/or stretched from the average distance of the nearest-neighbor Fe–Fe pairs [11].

In Fig. 3.70, the magnetic energy of the curve (a) has a sharp minimum and rapidly increases with m at a site with a small number of coordinate atoms contracted from the average nearest-neighbor distance [11]. The minimum position of the magnetic energy curve corresponds to the thermal average of the local moment $\langle m \rangle$ at each site in the ground state. In the paramagnetic state, the amplitude of the local moment $\langle m \rangle$ generally increases with temperature. However, the increase in $\langle m \rangle$ causes a rapid increase in the magnetic energy as shown by the curve (b), and hence the excitation to the small moment is preferred to decrease the free energy at finite temperatures. The excitation therefore results in a decrease in $\langle m^2 \rangle$ with increasing temperature. The number of excited states is limited, so that saturation of excitations takes place [122].

After saturation, $\langle m^2 \rangle$ begins to increase and the value of $\alpha_e + \alpha_m$ approaches a positive value. Such a saturation would also be established in the La(Fe$_{1-x}$Al$_x$)$_{13}$ amorphous alloys. On the other hand, the magnetic energy has a broad minimum or no minimum at finite m, when there are a large number of coordinate atoms on the contracted distance. At a site with such a magnetic energy, $\langle m^2 \rangle$ always increases in the paramagnetic temperature region [122]. There are various environments due to structural disorder, so that whether $\langle m^2 \rangle$ increases or decreases depends on the LEE of each site in amorphous alloys. Accordingly, $\alpha_e + \alpha_m$ gradually varies with temperature without a marked change and shows a minimum because of the balance between these positive and negative contributions. Such excitation, called a single-site excitation for SFs, contrasts with the collective excitation of SFs in WIE ferromagnets such as the $La(Ni_xAl_{1-x})_{13}$ amorphous alloys. From these discussions, it is considered that the thermal variation of $\alpha_e + \alpha_m$ is closely connected with the single-site excitation of SFs.

3.6 Magnetoelastic Properties and Linear Magnetostriction of $La(Fe_{1-x}Al_x)_{13}$ Amorphous Alloys

The remarkable thermal expansion anomaly in the $La(Fe_{1-x}Al_x)_{13}$ amorphous alloys shown in Fig. 3.64 is related to elastic properties. In this section, the ΔE effect, the linear saturation magnetostriction λ_S, the compressibility κ and the forced-volume magnetostriction $d\omega/dH$ are discussed in connection with the magnetovolume and magnetoelastic properties.

3.6.1 The ΔE Effect and Linear Magnetostriction

As is well known, the ΔE effect is associated with the magnetic domain, internal stress and so on, and generally common to ferromagnetic materials. It has been reported that Fe–Ni crystalline Invar alloys exhibit a pronounced elastic anomaly below T_C, even in the saturated magnetic field. The ΔE effect is described by three magnetic contributions using the following expression [142]:

$$\Delta E = \Delta E_\lambda + \Delta E_\omega + \Delta E_A = E_P - E , \qquad (3.29)$$

with

$$\Delta E_\lambda = -2\lambda_S E^2/5\sigma_i , \quad \Delta E_\omega = -E^2 \left(\frac{d\omega}{dH}\right)^2 \frac{1}{9\chi_{hf}} ,$$

and

$$\Delta E_A \approx -\omega_S ,$$

where ΔE_λ, ΔE_ω and ΔE_A are associated with the linear saturation magnetostriction λ_S, the forced-volume magnetostriction $d\omega/dH$, and the spontaneous volume magnetostriction ω_S, respectively. The Young's moduli E_P and E are the values in the paramagnetic state and in zero field, respectively. In (3.29), σ_i is the internal stress and χ_{hf} is the high-field susceptibility. Therefore, the anomalous thermal expansion given in Fig. 3.64 would bring about a pronounced anomalous temperature dependence in the Young's modulus.

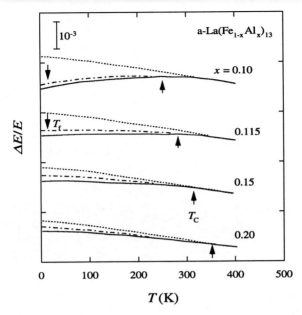

Fig. 3.71. Relative change in the Young's modulus vs. temperature in zero and 1.5 kOe magnetic fields for La(Fe$_{1-x}$Al$_x$)$_{13}$ amorphous alloys. The paramagnetic state is given by the *broken lines*. The data in zero and 1.5 kOe fields are given by the *solid* and *dash-dotted lines*, respectively [105]

Figure 3.71 depicts the temperature dependence of $\Delta E/E$ for the La(Fe$_{1-x}$Al$_x$)$_{13}$ amorphous alloys [36]. The solid and dash-dotted curves represent the results measured in magnetic fields of 0 kOe and 1.5 kOe, respectively. The external magnetic field of 1.5 kOe was applied in order to eliminate the magnetic domain effect. The temperature range of the paramagnetic state is too small to decide the slope of the elastic modulus. In order to estimate the paramagnetic Young's modulus, the temperature dependence of the Young's modulus for the Hf$_{40}$Fe$_{60}$ amorphous alloy was used as a reference (shown by the short-dashed curve), because this sample has a wide paramagnetic temperature range. The signs of the ΔE effect given by (3.29) are negative. As can be seen from the figure, the temperature dependence of the Young's modulus also indicates a marked anomaly below T_C, reflecting the thermal expansion anomaly. It should be noted that the Young's modulus measured in a magnetic field of 1.5 kOe is smaller than the paramagnetic Young's modulus, showing a significant softening with decreasing temperature. Such a peculiar phenomenon has also been observed for Zr–Fe [143] and Hf–Fe amorphous alloys [13]. The magnitude of softening increases markedly with increasing Fe content. This behavior is closely related to the Invar effect caused by the large spontaneous volume magnetostriction. More noticeable is the fact that the anomaly remains even well above T_C, as can be seen from Fig. 3.71. It

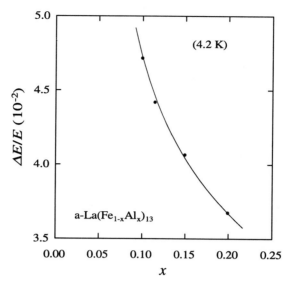

Fig. 3.72. Concentration dependence of the total ΔE effect, $\Delta E/E$, at 4.2 K for La(Fe$_{1-x}$Al$_x$)$_{13}$ amorphous alloys [105]

has been reported that the temperature dependence of the Young's modulus for Fe–Ni crystalline Invar alloys [144] and Fe$_{65}$(Ni$_{1-x}$Mn$_x$)$_{35}$ crystalline alloys [142] also exhibits a considerable softening even well above T_C.

It is clear that the La(Fe$_{1-x}$Al$_x$)$_{13}$ amorphous alloys exhibit softening in the shear mode over a wide temperature range because the shear modulus is given by $E/2(1 + v)$. The softening below and above T_C would be caused by the large spontaneous volume magnetostriction. It is interesting to note that a drastic increase in the Young's modulus for the above-mentioned crystalline alloys is observed at very low temperatures [142, 145], whereas the La(Fe$_x$Al$_{1-x}$)$_{13}$ amorphous alloys do not exhibit such an increase, as can be seen from Fig. 3.71. In Figs. 3.64 and 3.71, there is one further point to which we should draw attention, i.e., no distinct anomaly is observed in the thermal expansion and Young's modulus curves at the spin freezing temperature T_f. Figure 3.72 shows the concentration dependence of $\Delta E/E[= (E_P - E)/E]$ for the La(Fe$_{1-x}$Al$_x$)$_{13}$ amorphous alloys at 4.2 K. With increasing Fe content, $\Delta E/E$ increases in analogy with the high-field susceptibility χ_{hf} and the spontaneous volume magnetostriction $\omega_S(0)$ given in Figs. 3.42 and 3.65.

In order to see the effect of λ_S on the ΔE effect, the concentration dependence of $|\Delta E_\lambda|/E$ for La(Fe$_{1-x}$Al$_x$)$_{13}$ amorphous alloys at 4.2 K is given in Fig. 3.73, together with that of the squared magnetization M^2 in a magnetic field of 1.5 kOe at 4.2 K [36]. These curves have a broad maximum at around $x = 0.85$. The magnitude of $\Delta E_\lambda/E$ is based on the magnetic domains which contribute to the linear saturation magnetostriction λ_S. It is well known that, for Fe-based amorphous alloys, λ_S is associated with the magnetization M

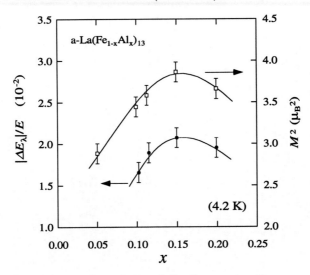

Fig. 3.73. Concentration dependence of the linear saturation magnetostriction term of the ΔE effect, $|\Delta E_\lambda|/E$, together with the square of the magnetization M^2 in 1.5 kOe at 4.2 K for La(Fe$_{1-x}$Al$_x$)$_{13}$ amorphous alloys [105]

and can be written [146]

$$\lambda_S \propto M^2 . \tag{3.30}$$

It is worth noting that the concentration dependence of the ratio $|\Delta E_\lambda|/E$ is similar to that of the squared magnetization measured at 1.5 kOe. Therefore, λ_S for the La(Fe$_{1-x}$Al$_x$)$_{13}$ amorphous alloys is expected to exhibit a similar concentration dependence (see Fig. 3.80).

The concentration dependence of $|\Delta E_\omega + \Delta E_A|/E$ associated with the forced-volume magnetostriction and the spontaneous volume magnetostriction for the La(Fe$_{1-x}$Al$_x$)$_{13}$ amorphous alloys is shown in Fig. 3.74 [36]. The value increases dramatically with decreasing x, or increasing Fe content. What has to be noticed is that the concentration dependence of $|\Delta E_\omega + \Delta E_A|/E$ exhibits a similar tendency to that of the spontaneous volume magnetostriction $\omega_S(0)$. Strictly speaking, the increase in $|\Delta E_\omega + \Delta E_A|/E$ is much more pronounced than that in $\omega_S(0)$ at low x, where the spin-glass state occurs [105]. In Fe-rich Zr–Fe amorphous alloys exhibiting spin-glass behavior at low temperatures, the forced-volume magnetostriction $d\omega/dH$ becomes very large [147]. This suggests that the ΔE_ω term would markedly increase at low x. It has been suggested that the La(Fe$_{1-x}$Al$_x$)$_{13}$ amorphous alloy system exhibits a large pressure effect on T_C because the spontaneous volume magnetostriction is directly reflected in the pressure effect on T_C. Furthermore, these alloys would also be expected to show a large compressibility due to the large magnetoelastic effects, as observed in other Invar-type amorphous alloys [13].

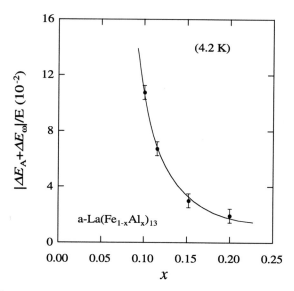

Fig. 3.74. Concentration dependence of the ΔE effect $|\Delta E_\omega + \Delta E_A|/E$ associated with the spontaneous volume magnetostrictions and forced-volume magnetostriction at 4.2 K for $La(Fe_{1-x}Al_x)_{13}$ amorphous alloys [105]

Figures 3.75 and 3.76 show the longitudinal and transverse linear magnetostrictions at 4.2 and 77 K, respectively, for four kinds of $La(Fe_{1-x}Al_x)_{13}$ amorphous alloy measured in magnetic fields up to 16 kOe [148]. The directions of the arrows in Fig. 3.75 indicate increasing and decreasing applied magnetic fields. We should not overlook the fact that the linear magnetostrictions of both directions $\lambda_\parallel$ and $\lambda_\perp$ at 4.2 K for the alloys with $x = 0.05$ and 0.10 do not return to the starting point when the applied magnetic field is lowered because of magnetic viscosity, characteristic of spin-glass alloys, whilst it disappears at 77 K as can be seen from Fig. 3.76. That is, the spin-glass state occurs at 4.2 K and the ferromagnetic state takes place at 77 K, in accord with the magnetic phase diagram Fig. 3.47.

Figure 3.77 shows the temperature dependence of the saturation magnetostriction λ_S for the ferromagnetic amorphous alloys with $x = 0.15$ and 0.20 [148]. These amorphous alloys exhibit a monotonic variation with temperature, but the RSG alloys with $x = 0.05$ and 0.10 show a broad maximum at the spin freezing temperature, as can be seen from Fig. 3.78. Shown in Fig. 3.79 is the temperature dependence of the offset $\Delta\lambda$ caused by the magnetic viscosity for the RSG alloys with $x = 0.05$ and 0.10. The value of $\Delta\lambda$ is defined by the value indicated in the schematic inset around the origin for the RSG in Fig. 3.47. The directions of the arrows indicate increasing and decreasing applied magnetic fields. The transverse magnetostriction $\lambda_\perp$ does not return to the starting point when the applied magnetic field is lowered to zero. The value of $\Delta\lambda$ disappears in the vicinity of 20 K for $x = 0.10$ and

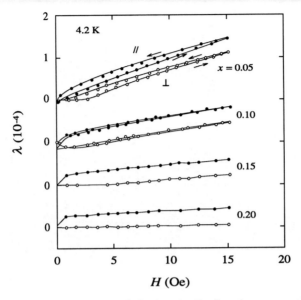

Fig. 3.75. The field dependence of the longitudinal and transverse linear magnetostriction, $\lambda_{\parallel}$ (-•-) and $\lambda_{\perp}$ (-o-), respectively, at 4.2 K for La(Fe$_{1-x}$Al$_x$)$_{13}$ amorphous alloys. Directions of *arrows* indicate increasing and decreasing applied magnetic fields [148]

70 K for $x = 0.05$, as can be seen from Fig. 3.79. It is worth noting that these temperatures correspond to their spin freezing temperature T_f in zero magnetic field (see Fig. 3.47). Accordingly, it is concluded that a clear magnetic viscosity is closely correlated to the SG state.

The concentration dependence of λ_S measured at 4.2 K in 10 kOe, together with that of the square of the magnetization M^2 measured in the same magnetic field for the La(Fe$_{1-x}$Al$_x$)$_{13}$ amorphous alloys [148] is plotted in Fig. 3.80. A maximum of about 26×10^{-6} is observed at $x = 0.15$, i.e., the value of λ_S increases with increasing Fe content within the ferromagnetic concentration range, whereas it decreases in the regime of the RSG state. The temperature dependence of λ_S has been accounted for by using classical one- or two-ion models [149–151] and the temperature is often converted to the corresponding magnetization. It has been pointed out that λ_S of Fe$_{1-x}$B$_x$ ($0.14 \leq x \leq 0.22$) and Fe$_{0.75}$P$_{0.15}$C$_{0.10}$ amorphous alloys is expected to scale as the square of the magnetization M^2 except at low temperatures [152,153], in accordance with the one-ion model [149,150]. We have discussed the data in Fig. 3.73 in the light of this model. Strictly speaking, however, the one-ion model seems to show a linear relationship in limited temperature ranges. The concentration dependence of M^2 measured in the same strength of applied magnetic field is also plotted in the same figure. The curve of λ_S exhibits a similar tendency to that of M^2, being consistent with (3.30). On the other

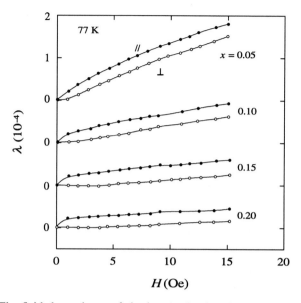

Fig. 3.76. The field dependence of the longitudinal and transverse linear magnetostriction, $\lambda_\parallel$ (-•-) and $\lambda_\perp$ (-○-), respectively, at 77 K for La(Fe$_{1-x}$Al$_x$)$_{13}$ amorphous alloys [148]

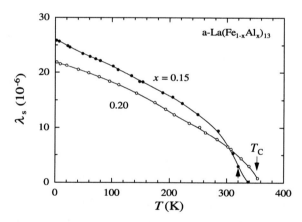

Fig. 3.77. Temperature dependence of the linear saturation magnetostriction λ_S for La(Fe$_{1-x}$Al$_x$)$_{13}$ ferromagnetic amorphous alloys with $x = 0.15$ and 0.20 [148]

hand, the two-ion model shows that λ_S is proportional to M^2 throughout the entire temperature range up to T_C.

Figure 3.81 shows the $\lambda_S(T)/\lambda_S(0)$ versus $M^2(T)/M^2(0)$ plots for the La(Fe$_{0.80}$Al$_{0.20}$)$_{13}$ amorphous ferromagnetic alloy in order to clarify the difference between these models. The results for the one-ion and two-ion models are given by dashed and solid lines, respectively. The data on the

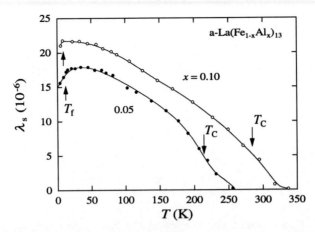

Fig. 3.78. Temperature dependence of the linear saturation magnetostriction λ_S for La(Fe$_{1-x}$Al$_x$)$_{13}$ re-entrant spin-glass amorphous alloys with $x = 0.05$ and 0.10 [148]

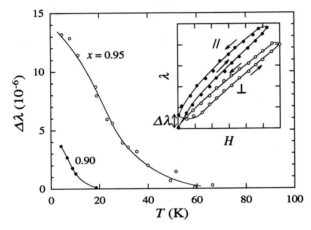

Fig. 3.79. Temperature dependence of the offset for the linear magnetostriction $\Delta\lambda$ due to the magnetic viscosity for La(Fe$_{0.90}$Al$_{0.10}$)$_{13}$ and La(Fe$_{0.95}$Al$_{0.05}$)$_{13}$ amorphous alloys [148]. The *inset* shows a schematic figure for defining the value of $\Delta\lambda$ in the vicinity of the origin for the re-entrant spin-glass (RSG) alloys given in Fig. 3.75. Directions of *arrows* indicate increasing and decreasing applied magnetic fields [148]

La(Fe$_{0.80}$Al$_{0.20}$)$_{13}$ amorphous alloy gives very nice plots on the straight solid line. Therefore, the results in Fig. 3.81 are explicable by the two-ion model [149, 150]. In NaZn$_{13}$-type La(Fe$_{1-x}$Al$_x$)$_{13}$ crystalline compounds, there are two Fe sites in the icosahedral clusters, namely, Fe^I and Fe^{II}, as shown in Fig. 3.4. The magnetic moments of these two sites are very different from each other in magnitude. That is to say, the value of Fe^I is $1.1\mu_B$ and that of Fe^{II} is $2.14\mu_B$ in the La(Fe$_{0.91}$Al$_{0.09}$)$_{13}$ crystalline compound [32]. The amor-

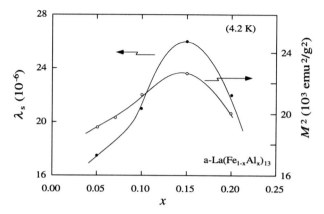

Fig. 3.80. Concentration dependence of the linear saturation magnetostriction λ_S at 4.2 K in 10 kOe, together with that of the square of the magnetization M^2 measured in the same magnetic field for $La(Fe_{1-x}Al_x)_{13}$ amorphous alloys [148]

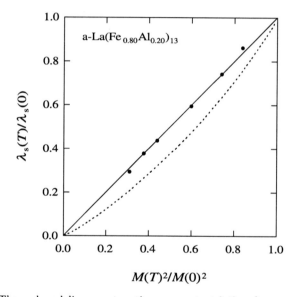

Fig. 3.81. The reduced linear saturation magnetostriction λ_S versus the square of the magnetization M^2 for the $La(Fe_{0.80}Al_{0.20})_{13}$ ferromagnetic amorphous alloy [148]. *Dashed* and *solid* lines stand for one-ion and two-ion models, respectively

phous counterparts also have icosahedral clusters composed of Fe^I and Fe^{II} atoms [24, 33]. Therefore, the magnetic state of Fe^I would be different from that of Fe^{II}, in support of the two-ion model.

3.6.2 Forced-Volume Magnetostriction and Compressibility

Large values of the spontaneous volume magnetostriction $\omega_S(T)$ and the high-field susceptibility χ_{hf} have been reported for La(Fe$_{1-x}$Al$_x$)$_{13}$ in the crystalline and amorphous states [34,36]. These results are highly relevant to the magnetovolume effects which give rise to Invar anomalies [12,111,154].

The temperature dependence of the giant forced-volume magnetostriction $d\omega/dH$ for Zr–Fe amorphous alloys has been investigated [147,155] and theoretically discussed using the Liberman–Pettifor formula, taking thermal SFs into account [6,137]. Because $d\omega/dH$ is proportional to the high-field susceptibility χ_{hf}, a large value of $d\omega/dH$ is expected in the La(Fe$_{1-x}$Al$_x$)$_{13}$ amorphous alloys. The magnetization of the La(Fe$_{1-x}$Al$_x$)$_{13}$ amorphous alloys is large, ranging from about 140 to 170 emu/g at 4.2 K [105]. Since large values of λ_S and $d\omega/dH$ are expected in the La(Fe$_{1-x}$Al$_x$)$_{13}$ amorphous alloys, the concentration and temperature dependences of λ_S and $d\omega/dH$ have been discussed. The magnetic phase diagram in the applied magnetic field has been obtained by measuring the differential magnetic susceptibility at various strengths of the magnetic field. These data will be connected with the spin-glass and ferromagnetic states and with a theoretical treatment related to itinerant-electron SGs [149,150].

During the last few decades, Brillouin scattering has been established as a powerful and convenient technique for determining both the magnetic constants and the elastic constants of metallic films [156,157]. This scattering is effective for determining the spin-wave dispersion coefficient of ferromagnetic thin films. In particular, for Invar-type materials, one can expect a marked softening of the Young's modulus and a low spin-wave dispersion coefficient. Therefore, Brillouin scattering can be a suitable technique for investigating the magnetic and elastic properties of Invar-type materials. The large compressibility of the Fe–B amorphous alloy has been confirmed by X-ray diffraction under pressure [158]. The compressibility obtained from the Brillouin scattering data for Ce–Fe amorphous alloys is extremely large [159]. A remarkable pressure effect on the electrical resistivity for Zr–Fe amorphous alloys has been correlated to their large compressibility [160]. The pressure effect on T_C for Zr–Fe [160], Hf–Fe [161], Sc–Fe [92], Nd–Fe–B [162,163] and La–Fe [112] amorphous alloys is significant. On the other hand, the pressure derivative of T_C for various amorphous Fe-based alloys has been estimated from $d\omega/dH$ and these data have been connected with the magnetic inhomogeneity [164].

Figure 3.82 shows the temperature dependence of the forced-volume magnetostriction $d\omega/dH$ for the ferromagnetic La(Fe$_{1-x}$Al$_x$)$_{13}$ amorphous alloys with $x = 0.15$ and 0.20 [148]. The value of $d\omega/dH$ is obtained from

$$\frac{d\omega}{dH} = \left(\frac{d\lambda}{dH}\right)_{\parallel} + 2\left(\frac{d\lambda}{dH}\right)_{\perp} , \qquad (3.31)$$

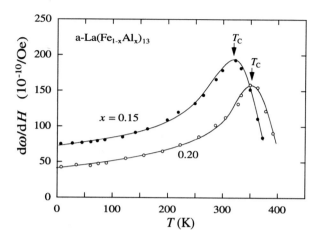

Fig. 3.82. Temperature dependence of the forced-volume magnetostriction dω/dH for ferromagnetic La(Fe$_{1-x}$Al$_x$)$_{13}$ amorphous alloys with $x = 0.15$ and 0.20 [148]

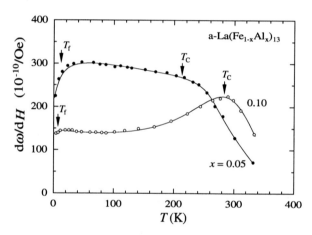

Fig. 3.83. Temperature dependence of the forced-volume magnetostriction dω/dH for re-entrant spin-glass (RSG) La(Fe$_{1-x}$Al$_x$)$_{13}$ amorphous alloys with $x = 0.05$ and 0.10 [148]

where $(d\lambda/dH)_{\parallel}$ and $(d\lambda/dH)_{\perp}$ are the longitudinal and transverse forced linear magnetostriction, respectively. These curves exhibit a peak at T_C. The temperature dependence of dω/dH for the La(Fe$_{1-x}$Al$_x$)$_{13}$ RSG amorphous alloys with $x = 0.05$ and 0.10 is also given in Fig. 3.83 [148]. The magnitude of dω/dH for both ferromagnetic and RSG amorphous alloys is very large and comparable to that of Zr–Fe and Zr–Fe–Ni amorphous alloys [147, 155]. The value for the La(Fe$_{0.95}$Al$_{0.05}$)$_{13}$ amorphous alloy is about twice that for Fe–Ni Invar alloy and Zr(Fe$_{1-x}$Co$_x$)$_2$ crystalline compounds [165]. Note that dω/dH of α (bcc) Fe and γ (fcc) Ni is 5×10^{-10} and 1×10^{-10} Oe^{-1}, respectively,

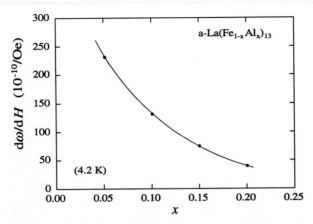

Fig. 3.84. Concentration dependence of the forced-volume magnetostriction $d\omega/dH$ at 4.2 K for La(Fe$_{1-x}$Al$_x$)$_{13}$ amorphous alloys [148]

at room temperature [166, 167]. The transition temperature T_f is affected by applying a magnetic field, as can be seen from Fig. 3.53, but the variation of T_f is not so significant in a magnetic field above several hundred Oe. On the other hand, T_C is slightly increased as mentioned in Fig. 3.54. Accordingly, we can define T_C and T_f as given by the arrows in Figs. 3.82 and 3.83. The peak at T_C is reduced with decreasing x and eventually disappears in the curve of $x = 0.05$.

The forced-volume magnetostriction $d\omega/dH$ in the RSG regime around pure Fe in the amorphous state has been investigated on the basis of the finite-temperature theory of LEEs for amorphous metallic magnetism [6,137]. According to this model, the peak at the freezing temperature is caused by the change in the amplitudes of local magnetic moments, which is characteristic of itinerant-electron SGs. In contrast, such a peak is hardly observed in SGs in localized electron magnetic moment systems [6]. Furthermore, the peak corresponding to T_f becomes clear with decreasing T_C, whereas the peak at T_C becomes obscure. The data in Fig. 3.83 seem to be consistent with this theoretical discussion mentioned above [137], although the freezings are similar to those derived from the Ising or Heisenberg model [168,169].

The concentration dependence of $d\omega/dH$ at 4.2 K for La(Fe$_{1-x}$Al$_x$)$_{13}$ amorphous alloys is illustrated in Fig. 3.84 [148]. As is well known, the conventional expression for the forced-volume magnetostriction $d\omega/dH$ is

$$\frac{d\omega}{dH} = 2\kappa C M \frac{dM}{dH} = 2\kappa C M \chi_{hf} , \qquad (3.32)$$

where $dM/dH = \chi_{hf}$ is the high-field susceptibility. On the other hand, in a recent theoretical model taking thermal SFs into consideration [170], $d\omega/dH$ is calculated from [6,137]

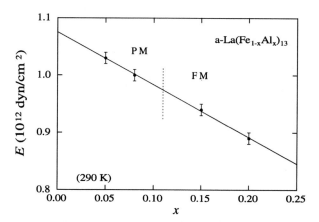

Fig. 3.85. Concentration dependence of the room temperature value of the Young's modulus E determined by Brillouin scattering data for $La(Fe_{1-x}Al_x)_{13}$ amorphous alloys [114]

$$\frac{d\omega}{dH} = \frac{\Gamma_e\kappa}{3V}\left(T\frac{d[\langle m_z\rangle]_S}{dT} + \frac{1}{4}U\frac{d[\langle m^2\rangle]_S}{dH}\right),\qquad(3.33)$$

where $\langle\ \rangle$ and $[\ ]_S$ denote the thermal and structural averages, respectively. It has been pointed out that the second term in (3.33) is dominant in itinerant-electron SGs [6,137]. From (3.32) and (3.33), $d\omega/dH$ is proportional to the compressibility κ, which is obtained from the conventional expression

$$\kappa = 3(1-2v)/E,\qquad(3.34)$$

where v is Poisson's ratio and E is the Young's modulus. These two values at room temperature have been determined from the surface acoustic velocities of the Rayleigh and Sezawa waves in Brillouin scattering data.

The concentration dependence of the room temperature Young's modulus E is given in Fig. 3.85 [114]. Furthermore, the relative value of E against temperature is available from Fig. 3.71. Therefore, the temperature dependence of E for the $La(Fe_{1-x}Al_x)_{13}$ amorphous alloys can be obtained [36]. The value of E is much lower than that of a crystalline α (bcc) Fe and about 0.9–1.0×10^{12} dyn/cm^2, depending on the concentration. The value of v is very close to $1/3$, and hence $\kappa \approx 1/E$. Consequently, we can obtain the compressibility κ at 4.2 K from these results for the $La(Fe_{1-x}Al_x)_{13}$ amorphous alloys. The value of κ thus obtained is plotted against the concentration x in Fig. 3.86 [171]. The concentration dependence is similar to that of various other magnetovolume and magnetoelastic properties (see Figs. 3.42, 3.65 and 3.84). The value of κ is very large, ranging from 0.88 to 0.98×10^{-12} cm^2/dyn in the range $0.10 \leq x \leq 0.20$, being about twice that of the crystalline α (bcc) Fe. As a consequence, the large values of $d\omega/dH$ in Figs. 3.84 are closely correlated to this large value of κ. These results on $d\omega/dH$, χ_{hf}, κ and $\omega_S(T)$

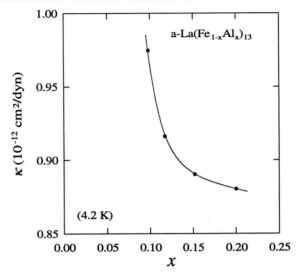

Fig. 3.86. Concentration dependence of the compressibility κ at 4.2 K for La(Fe$_{1-x}$Al$_x$)$_{13}$ amorphous alloys [35]

exhibit a similar trend, i.e., a remarkable increase occurs with decreasing x or increasing Fe. Consequently, these physical properties are closely interrelated.

3.7 Effects of Pressure on the Magnetic Properties of La(Fe$_{1-x}$Al$_x$)$_{13}$ Amorphous Alloys

The experimental and theoretical results of the pressure effect on T_C of WIE ferromagnets in both crystalline and amorphous states have recently been discussed in detail [65]. A general relation between the pressure effect on T_C and M has been derived [172]. More recently, the effects of pressure on the magnetic properties of amorphous Fe have been investigated on the basis of a finite-temperature theory of LEEs [6]. A number of data on the pressure effect on T_C have been reported. However, studies on the effect of pressure on M for amorphous alloys are not so active as those for crystalline alloys [173, 174], although they would be expected to give valuable information.

A pressure-induced transition from the ferromagnetic to the antiferromagnetic state has been observed at a pressure of $P \leq 0.1$ GPa in the La(Fe$_{0.86}$Al$_{0.14}$)$_{13}$ crystalline compound [107]. From high pressure experiments of the Mössbauer effect on magnetic properties of La(Fe$_{0.88}$Al$_{0.12}$)$_{13}$, it has been pointed out that the Curie temperature and the average magnetic hyperfine field decrease abruptly at a critical pressure corresponding to the average Fe–Fe nearest-neighbor distance $r_C = 0.253$ nm [175]. Significant effects of pressure on T_C and T_N for La(Fe$_{0.88-x}$Co$_x$Al$_{0.12}$)$_{13}$ crystalline

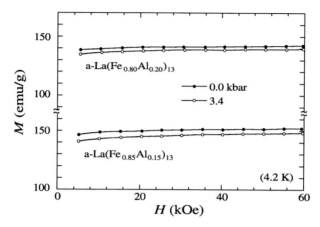

Fig. 3.87. Magnetization curves up to 60 kOe at 4.2 K under different pressures for La(Fe$_{1-x}$Al$_x$)$_{13}$ ($x = 0.15$ and 0.20) amorphous alloys [171]

compounds have been discussed in the context of the instability of ferromagnetism [176]. Judging from the above, marked pressure effects are expected on the magnetic properties of the La(Fe$_{1-x}$Al$_x$)$_{13}$ amorphous alloys, because the Fe–Fe interatomic distance becomes short due to the large compressibility. Therefore, the pressure effects on the magnetization, the Curie temperature and the spin freezing temperature have been discussed. These data are compared with theoretical calculations for pure Fe in the amorphous state [6], and discussed in connection with the large magnetovolume effects and magnetic instability.

3.7.1 Effect of Pressure on Magnetization

As explained in the preceding sections, the La(Fe$_{1-x}$Al$_x$)$_{13}$ amorphous alloys exhibit various remarkable magnetovolume effects in analogy with the crystalline counterparts [34]. Therefore, significant pressure effects on the magnetic properties are expected.

Figures 3.87 and 3.88 show the magnetization curves up to 60 kOe under pressure for the La(Fe$_{1-x}$Al$_x$)$_{13}$ ferromagnetic amorphous alloys with $x = 0.15$ and 0.20 and those for the RSG alloys with $x = 0.05$ and 0.10, respectively. The magnetization decreases with increasing applied pressure, and the magnitude of the decrease in the magnetization depends on the magnetic state mentioned above. In more detail, the decrease for ferromagnetic amorphous alloys with $x = 0.15$ and 0.20 is not so remarkable, whereas the RSG amorphous alloys with $x = 0.90$ and 0.95 exhibit a significant decrease. The pressure dependence of M measured at 4.2 K in 60 kOe is plotted in Fig. 3.89 [171]. The decrease in M becomes significant with increasing x. The concentration dependence of dM/dP is given in Fig. 3.90. The value of |dM/dP| drastically increases in the RSG concentration regime. It should

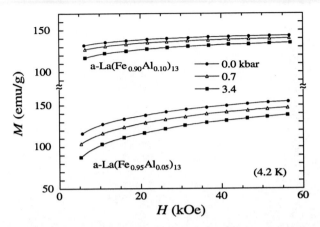

Fig. 3.88. Magnetization curves up to 60 kOe at 4.2 K under different pressures for La(Fe$_{1-x}$Al$_x$)$_{13}$ ($x = 0.05$ and 0.10) amorphous alloys [171]

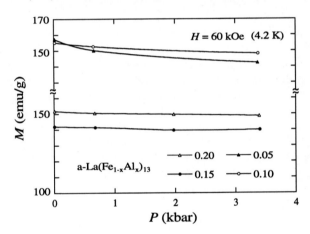

Fig. 3.89. Pressure dependence of the magnetization measured at 4.2 K in 60 kOe for La(Fe$_{1-x}$Al$_x$)$_{13}$ amorphous alloys [171]

be noted that the spontaneous volume magnetostriction ω_s and the forced-volume magnetostriction $d\omega/dH$ also drastically increase with decreasing x [171]. From Maxwell's relation for the free energy, the pressure effect on the magnetization is equivalent to $d\omega/dH$, and given by

$$\frac{d\omega}{dH} = -\rho_m \frac{dM}{dP} , \qquad (3.35)$$

where ρ_m is the mass density. Therefore, the concentration dependence of $|dM/dP|$ should be similar to that of $d\omega/dH$. The magnetic properties of Fe are predominantly governed by environmental factors such as the coordination number CN and the Fe–Fe interatomic distance r. Various theoretical

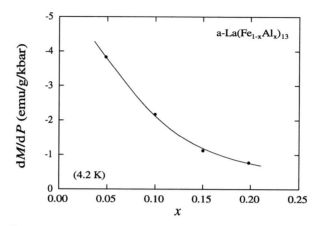

Fig. 3.90. Concentration dependence of the pressure derivative of magnetization dM/dP at $4.2\,K$ for $La(Fe_{1-x}Al_x)_{13}$ amorphous alloys [171]

and experimental results tell us that the drastic change in the magnetism from the antiferromagnetic to the ferromagnetic state occurs near an Fe–Fe distance of 0.25 nm, or a lattice constant of 0.35 nm, in γ (fcc) Fe [111], as shown in Fig. 3.39. In fact, the average magnetic hyperfine field of Fe–Ni crystalline alloys shows a marked decrease below 0.25 nm [107]. The influence of pressure on the Mössbauer effect has been studied in the $La(Fe_{0.88}Al_{0.12})_{13}$ crystalline compound. An abrupt change occurs in the average magnetic hyperfine field at a critical pressure of about 4.5 GPa, corresponding to the Fe–Fe distance of about 0.253 nm [175]. Icosahedral clusters are composed of Fe and Al atoms and the CN value of Fe depends on x, but the Fe–Fe nearest-neighbor interatomic distance scarcely changes with x, being about 0.255 nm. With increasing Fe content, therefore, the ferromagnetism becomes unstable along with the increase in the nearest-neighbor Fe atoms, and hence the RSG state is induced, resulting in the remarkable pressure effect mentioned above.

Under hydrostatic pressure, the ferromagnetic amorphous alloys with $x = 0.15$ and 0.20 exhibit a linear relation between the magnetization $M(T)$ measured at 10 kOe and $T^{3/2}$ over a wide range of temperatures, as shown in Fig. 3.91, being dominated by spin-wave excitations given by the Bloch formula (3.6). The obtained spin-wave dispersion coefficient D_M is 48.5 for $x = 0.15$ and $51.5\,meV\,\text{Å}^2$ for $x = 0.20$. From the relation M versus $T^{3/2}$ in Fig. 3.91, the spin-wave dispersion coefficient D_M is obtained for the ferromagnetic alloys with $x = 0.15$ and 0.20. These values are much smaller than those of Fe–metalloid amorphous alloys [113]. On the other hand, the RSG amorphous alloys with $x = 0.10$ and 0.05 do not obey the $T^{3/2}$ law, but the magnetization M exhibits a linear dependence on T^2 over wide temperature ranges due to the Stoner-type excitation in WIE ferromagnets, as shown in Fig. 3.92, even though the Stoner theory does not correctly predict the finite temperature ferromagnetism [177]. It should be pointed out that

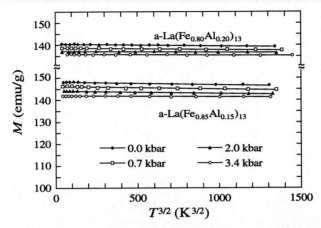

Fig. 3.91. Temperature dependence of magnetization in the form M–$T^{3/2}$ under different pressures for ferromagnetic La(Fe$_{1-x}$Al$_x$)$_{13}$ amorphous alloys with $x = 0.15$ and 0.20 [171]

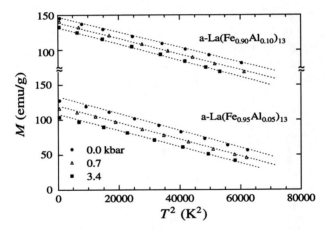

Fig. 3.92. Temperature dependence of magnetization in the form M–T^2 under different pressures for re-entrant spin-glass La(Fe$_{1-x}$Al$_x$)$_{13}$ amorphous alloys with $x = 0.05$ and 0.10 [171]

the deviation from the straight line in the low temperature ranges is due to the spin-glass state. Comparing the figures under such pressures, it is considered that the spin-glass state is not created by reducing T_C for alloys with $x = 0.15$ and 0.20, because no deviation is confirmed at low temperatures in Fig. 3.91, in contrast to Fig. 3.92.

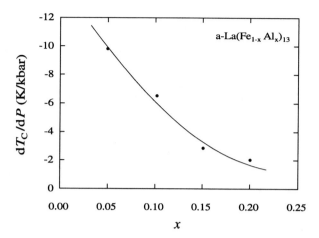

Fig. 3.93. Concentration dependence of the pressure derivative of the Curie temperature dT_C/dP of La(Fe$_{1-x}$Al$_x$)$_{13}$ amorphous alloys [171]

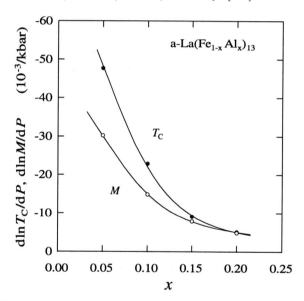

Fig. 3.94. Concentration dependence of the pressure coefficient of the Curie temperature $d\ln T_C/dP$ and the magnetization $d\ln M/dP$ for La(Fe$_{1-x}$Al$_x$)$_{13}$ amorphous alloys [171]

3.7.2 Effect of Pressure on the Curie Temperature

For WIE ferromagnets in the homogeneous state, the relation between $d\ln T_C/dP$ and $d\ln M/dP$ has been obtained by expanding the free energy as a power series in the magnetization [65]. In the Stoner–Wohlfarth model, the pressure coefficients are given by

$$\frac{d \ln T_C}{dP} = \frac{d \ln M}{dP} + \frac{6}{5}\kappa + \frac{1}{2}\frac{d \ln b_L}{dP} , \tag{3.36}$$

where b_L is the Landau expansion coefficient of the fourth power of the magnetization. On the other hand, the following general relation for the pressure effects mentioned above has been obtained [172] by making use of the Lang–Ehrenreich model [178] without using the expansion of the free energy:

$$\frac{d \ln T_C}{dP} = \frac{5}{3}\kappa + \frac{C_{eff}}{\chi_{hf}T_C}\frac{d \ln M}{dP} , \tag{3.37}$$

where C_{eff} is the effective Curie constant. In the WIE ferromagnetic limit, i.e., $C_{eff}/\chi_{hf}T_C$ equal to unity, this equation reduces to the expression obtained by expansion. A study of $C_{eff}/\chi_{hf}T_C$ for Fe–Ni crystalline and Fe$_{90-x}$Ni$_x$Zr$_{10}$ amorphous alloys has been carried out and it has been pointed out that the $C_{eff}/\chi_{hf}T_C$ versus T_C plots for both alloy systems give unity at about $T_C = 290$ K. It should be noted that the Curie temperature of the La(Fe$_{1-x}$Al$_x$)$_{13}$ amorphous alloy with $x = 0.10$ is close to this temperature, as can be seen from Fig. 3.47 [105]. Figure 3.93 shows the concentration dependence of dT_C/dP determined from the thermomagnetization curves for the La(Fe$_{1-x}$Al$_x$)$_{13}$ amorphous alloys [171]. The magnitude of dT_C/dP significantly increases with decreasing x in analogy with χ_{hf}, $d\omega/dH$ and dM/dP, which are closely correlated to ω_S (see Figs. 3.42, 3.84 and 3.90). The Curie temperature differs from the magnetization in the pressure coefficient, i.e.,

$$\frac{d \ln T_C}{dP} < \frac{d \ln M}{dP} . \tag{3.38}$$

This relation has been reported for Fe$_{65}$(Ni$_{1-x}$Mn$_x$)$_{35}$ crystalline alloys [174], non-stoichiometric Ni$_3$Al compounds [173] and Ni–Cr crystalline alloys [179]. Such behavior could be explained by considering a large compressibility κ [164] and, in addition, the positive Landau expansion coefficient obtained from the Arrott plot [64]. On the other hand, some weak ferromagnetic compounds [180], Zr$_{10}$(Fe$_{1-x}$Ni$_x$)$_{90}$ amorphous alloys [181, 182] and the band calculation for amorphous Fe [6] exhibit an opposite relation in magnitude, depending on the alloy composition and/or input parameters. As indicated in Fig. 3.94, the results for the La(Fe$_{1-x}$Al$_x$)$_{13}$ amorphous alloys fall within the latter case. It is noted that the SF theory gives $d\ln T_C/d\ln M = 3/2$ for WIE ferromagnets [110].

The magnitude of $\omega_S(T)$ is predominantly governed by κC. Here, κ is the compressibility and C is the coupling constant defined in terms of the Landau theory of phase transitions [183,184]. The values of ω_S for the La(Fe$_{1-x}$Al$_x$)$_{13}$ amorphous alloys are extremely large [36], and hence κC is also large [148]. The marked effects of pressure on the magnetization and the Curie temperature are also closely correlated to the large compressibility κ. The magnitude of dT_C/dP is proportional to κC and χ_0 and given by [65]

$$\frac{dT_C}{dP} = 2\kappa C \chi_0 T_C , \tag{3.39}$$

where χ_0 is the zero-field susceptibility given by

$$\chi_0 = \mu_B \rho(\varepsilon_F) \frac{T_F^2}{T_C^2} . \qquad (3.40)$$

Here $\rho(\varepsilon_F)$ is the DOS at the Fermi level, μ_B is the Bohr magneton, and T_F is the effective degeneracy temperature. The values of κ for the $La(Fe_{1-x}Al_x)_{13}$ amorphous alloys are significantly large (see Fig. 3.86). Since κC is associated with $\omega_S(T)$, the ΔE effect and $d\omega/dH$ show a similar concentration dependence (see Figs. 3.72 and 3.84). The pressure effect has also been discussed, taking SFs into consideration, and given by [185]

$$\frac{dT_C}{dP} \propto -2\kappa C\rho(\varepsilon_F)\mu_B^2 \frac{T_F^{4/3}}{T_C^{1/3}} . \qquad (3.41)$$

As can be seen from Fig. 3.95, the dT_C/dP versus $T_C^{-1/3}$ plot gives a relatively nice linear relationship for the $La(Fe_{1-x}Al_x)_{13}$ amorphous alloys. The plot for Zr–Fe amorphous alloys is given in the same figure for comparison [161]. As can be seen from the figure, the slope of the $La(Fe_{1-x}Al_x)_{13}$ amorphous alloys is about twice that of the Zr–Fe amorphous alloys. Since the spontaneous volume magnetostriction ω_S of these two alloys is not so different [13,36], the magnitude of κC would be comparable. The DOS may not be so different because the ferromagnetism of both these alloy systems is not very stable. Consequently, the different slope may be mainly attributed to the different effective degeneracy temperature which is correlated with the first and second derivatives of the DOS at the Fermi level ε_F. In other words, the difference between the $La(Fe_{1-x}Al_x)_{13}$ and Zr–Fe amorphous alloys for the structural and concentration fluctuations would give the different values of T_F.

The amorphous structures can influence both the DOS and the electron interactions, and hence the effects of amorphicity should be taken into consideration. The LG formalism concerned with concentration fluctuation has been tried as a way of discussing pressure effects [65]. It has been pointed out that, empirically, the $d\ln T_C/dP$ vs. T_C plot becomes linear for magnetically inhomogeneous systems. Substituting $\chi_1 - \chi_2 T_C$ for χ_0 in (3.40), the following empirical expression yields the spin-glass-like peak for $T_C = 0$ [65]:

$$\frac{d\ln T_C}{dP} = \frac{1}{T_C}\frac{dT_C}{dP} = -j + \kappa T_C , \qquad (3.42)$$

with $j = 2\kappa C\chi_1$ and $k = 2\kappa C\chi_2$. Figure 3.96 shows the plot in the form of (3.42) for the present amorphous alloys, together with that for the Zr–Fe amorphous alloys [161] for comparison. A linear relationship is observed, although the data show some scatter.

Since the bulk modulus $B = \kappa^{-1}$ is obtained from the second volume derivative of the total energy, the notation BdM/dP and BdT_C/dP is physically more meaningful. The concentration dependences of these two pressure

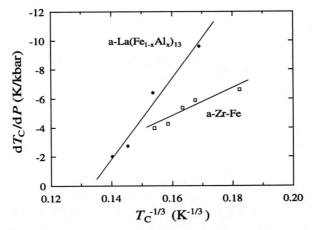

Fig. 3.95. Pressure derivative of the Curie temperature dT_C/dP vs. $1/T_C^{1/3}$ plot for La(Fe$_{1-x}$Al$_x$)$_{13}$ amorphous alloys, together with that for Fe–Zr amorphous alloys [171]

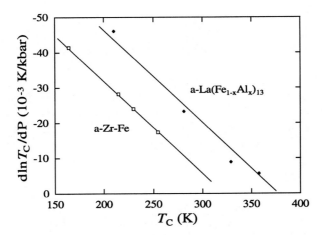

Fig. 3.96. Pressure coefficient of the Curie temperature $d\ln T_C/dP$ vs. T_C plots for La(Fe$_{1-x}$Al$_x$)$_{13}$ amorphous alloys [171], together with that for Fe–Zr amorphous alloys [82]

derivatives are very similar [171]. With decreasing x, these two pressure effects become significant in accordance with the instability of ferromagnetism. Taking into account both thermal SFs and local magnetic moment fluctuations with respect to the structural disorder, the theoretical investigations also reveal that the large values of these two pressure effects are associated with the instability of ferromagnetism [6].

Finally, we turn to the effect of pressure on the RSG state. For the La(Fe$_{0.90}$Al$_{0.10}$)$_{13}$ amorphous alloy, the effect of pressure in 100 Oe is given in

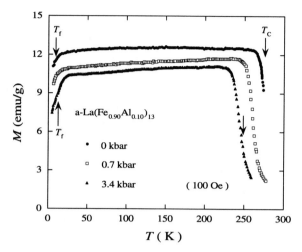

Fig. 3.97. Temperature dependence of the magnetization in a field of 100 Oe as a function of hydrostatic pressure for the La(Fe$_{0.90}$Al$_{0.10}$)$_{13}$ amorphous alloy [171]

Fig. 3.97. The data on T_C and T_f are obtained from the magnetic cooling effect in 100 Oe using the same clamp pressure cell for the magnetization measurements. By applying pressure, T_f is increased, whereas T_C is reduced, as can be seen from Fig. 3.97. Note that theoretical analysis for itinerant-electron SGs taking SFs into consideration gives a consistent explanation [6]. Since the spin freezing temperature T_f is very sensitive to the magnetic field (see Fig. 3.53), it should be borne in mind that the present results are different from those reported before because of the different experimental conditions [105].

The dependences of T_C and T_f on pressure have been obtained numerically by considering the derivative dT_C/dV for Fe-based amorphous alloys [6]. The calculated T_C decreases with increasing bandwidth W, while T_f increases until the ferromagnetism disappears at a critical pressure. These calculated results are consistent with the experimental results for La–Fe amorphous alloys [112]. Note that the shifts for the La(Fe$_{0.95}$Al$_{0.05}$)$_{13}$ amorphous alloy are more significant than those of the La(Fe$_{0.90}$Al$_{0.10}$)$_{13}$ amorphous alloy, reflecting the magnetic instability (see Fig. 3.47).

3.8 Concluding Remarks

We have discussed atomic structures, SFs, SG behavior, magnetovolume, and pressure effects in the La(TM$_{1-x}$Al$_x$)$_{13}$ amorphous alloys composed of icosahedral clusters.

The composition ranges in the amorphous phase are extended, compared with their counterparts in the crystalline phases. Furthermore, the phase composed of the icosahedral clusters is formed even in systems having no

NaZn$_{13}$-type crystalline phase. The local amorphous structures have been compared with those of NaZn$_{13}$-type crystalline compounds. The influence of thermal variation on SFs has been observed in the thermal variation of magnetization below and above the Curie temperature T_C, and in the magnetization process at 4.2 K.

The structural parameters obtained by X-ray diffraction data strongly imply that the local environment of the La(TM$_{1-x}$Al$_x$)$_{13}$ amorphous alloys consists of icosahedral clusters. The La(Ni$_{1-x}$Al$_x$)$_{13}$ amorphous alloys are considered to be good magnetic materials because these alloys are classified as WIE ferromagnets. The temperature dependence of magnetization for La(Ni$_{1-x}$Al$_x$)$_{13}$ exhibits an M–$T^{3/2}$ relation associated with the spin-wave excitation at low temperatures. The temperature dependence of the magnetization exhibits an M^2–$T^{4/3}$ relation below the Curie temperature T_C. This relation holds well in lower Ni concentration ranges. With increasing Ni concentration, the M^2–$T^{4/3}$ relation becomes poor, whereas an M^2–T^2 relation appears over a wide temperature range. The inverse magnetic susceptibility χ^{-1} vs. T curve is convex upwards just above T_C and exhibits a remarkable upturn in slope at high temperatures due to the saturation of thermal growth of SFs. In low Ni concentration regions, the generalized Rhodes–Wohlfarth plots indicate that the local spin density is thermally variable and its fluctuations dominate the characteristics of the χ^{-1}–T curve.

In Co systems such as LaCo$_{13}$ and La(Co$_{1-x}$Al$_x$)$_{13}$, ferromagnetism is enhanced and SF features come close to the localized moment limit. Hence, no remarkable SF behavior is observed. In the La(Co$_{1-x}$Mn$_x$)$_{13}$ amorphous alloys, the concentration dependence of the magnetic moment exhibits a broad maximum due to the virtual bound state in the DOS.

Since the La(Fe$_{1-x}$Al$_x$)$_{13}$ amorphous alloys are situated at the intermediate state in the SF features, many difficulties arise in detailed discussions. This alloy system exhibits an RSG behavior in the concentration range $0.05 \leq x \leq 0.15$, where antiferromagnetic order takes place in the crystalline state. The spin-glass state is responsible for a significant high-field susceptibility at 4.2 K. The spin freezing temperature hardly depends on the kind of RE in accordance with the unchanged structural parameters of icosahedral clusters. The thermal expansion anomaly due to a remarkably large volume magnetostriction $\omega_S(T)$ is observed in a wide temperature range, and it becomes larger with increasing Fe content. The thermal expansion anomaly above the Curie temperature T_C is closely associated with a large contraction in the amplitude of the local moment. The temperature dependence of the Young's modulus shows a marked anomaly, corresponding to the thermal expansion anomaly. The anomaly becomes significant with increasing Fe content and remains even well above the Curie temperature. The pressure coefficient of the Curie temperature $d\ln T_C/dP$ is explained in terms of the SFs and/or the magnetic inhomogeneity.

On these grounds, we can say with a fair degree of certainty that the various magnetic properties of the La(TM$_{1-x}$Al$_x$)$_{13}$ amorphous alloys composed of icosahedral clusters with well defined local atomic structures are closely correlated to the SFs, although there is room for argument concerning magnetic inhomogeneity in the amorphous structures.

Acknowledgements

The authors are grateful to Professor T. Goto of the University of Tokyo, Professor H. Tange of Ehime University, and Professor A. Yoshihara of Ishinomaki Senshu University for their effective collaboration and helpful discussions.

List of Abbreviations and Symbols

CN	coordination number
DE	Damon–Eshbach
DOS	density of states
EW	Edward–Wohlfarth
FC	field-cooled
FM	ferromagnetic
FSM	fixed-spin moment
GRW	generalized Rhodes–Wohlfarth
HF	Hartree–Fock
IS	isomer shift
LEE	local environment effect
LG	Landau–Ginzburg
LMTO	linear muffin-tin orbital
MK	Moriya–Kawabata
PM	paramagnetic
QS	quadrupole splitting
RDF	radial distribution function
RE	rare earth element
RIF	reduced interference function
RPA	random phase approximation
RSG	re-entrant spin glass
RW	Rhodes–Wohlfarth
SCR	self-consistent renormalization
SF	spin fluctuation
SG	spin glass
SRO	short-range ordering
SSW	standing spin wave
TM	transition metal
VBS	virtual bound state

WIE	weak itinerant-electron
ZFC	zero-field-cooled
A	exchange stiffness constant
B	bulk modulus
C	magnetoelastic coupling constant
C_{eff}	effective Curie constant
C_{V}	specific heat
C_{V}^{e}	electronic specific heat
D	spin-wave dispersion coefficient
$\mathrm{d}\omega/\mathrm{d}H$	forced-volume magnetostriction
E	Young's modulus
H	magnetic field
I	Coulomb interaction per electron
k_{B}	Boltzmann constant
K_q	magnetovolume coupling constant for the SF component with q
m	local moment
M	magnetization
M_{B}	bulk magnetization
m_{e}	mass of electron
M_{S}	surface magnetization
M_{SP}	spontaneous magnetization
M_{ST}	saturation magnetization
M_{TM}	magnetic moment of $3d$ transition metal (TM = Fe, Co, Ni, Mn)
N	total number of constituent elements
N_{A}	number of atoms per unit volume
N_d	d electron number
N_{e}	total electron number
N_0	total number of magnetic atoms
P	pressure
P_{C}	magnetic carrier number
P_{eff}	effective moment
P_{S}	saturation moment
q	wavenumber of magnetic excitations
Q	X-ray scattering wave vector
$Qi(Q)$	reduced interference function
r	interatomic distance
R	distance
S_{L}^2	mean-squared local amplitude of spin fluctuations
S_q	spin density for the SF component with q
T	temperature
T^*	saturation temperature of S_{L}^2
T_{C}	Curie temperature
T_{f}	spin freezing temperature of the longitudinal component

T_F	effective degeneracy temperature
T_g	spin freezing temperature of the transverse component
T_N	Néel temperature
U	effective Coulomb energy
V	bulk volume
x	concentration

α	thermal expansion coefficient
$\beta(T)$	specific coefficient of the phonon term
Γ	Grüneisen parameter
Γ_e	electronic Grüneisen parameter
γ	gyromagnetic ratio
ε	energy level
ε_F	Fermi level
θ_D	Debye temperature
κ	compressibility
λ	X-ray wavelength
λ_0	vacuum wavelength of laser light
λ_S	linear saturation magnetostriction
μ_B	Bohr magneton
$\rho(\varepsilon_F)$	density of states at the Fermi surface
ρ_m	room temperature density
ρ_0	average number density
σ_i	internal stress
υ	Poisson's ratio
χ	susceptibility
χ_0	zero-field susceptibility
χ_{hf}	high-field susceptibility
Ω	volume strain
ω	frequency of magnetic excitations
ω_R	resonance frequency of Brillouin scattering
$\omega_s(T)$	spontaneous volume magnetostriction at T
Ψ	Stoner enhancement factor

References

1. F. Bloch: Z. Physik **57**, 545 (1929)
2. T. Moriya: *Spin Fluctuation in Itinerant Electron Magnetism* (Springer-Verlag, Berlin, 1985)
3. E.C. Stoner: Proc. Roy. Soc. A **165**, 372 (1938)
4. K. Yoshida: *Theory of Magnetism* (Springer-Verlag, Berlin, 1996)
5. T. Fujiwara: J. Non-Cryst. Solids **61–62**, 1039 (1984)
6. Y. Kakehashi: Phys. Rev. B **47**, 3185 (1993)

7. Y. Kakehashi, T. Uchida and M. Yu: Phys. Rev. B **56**, 8807 (1997)
8. A.M. Bratkovsky and A.V. Smirnov: J. Phys. Condens. Matter **5**, 3203 (1993)
9. W.Y. Ching, G.L. Zhao and Y. He: Phys. Rev. B **42**, 10878 (1990)
10. H. Tanaka and S. Takayama: J. Phys. Condens. Matter **4**, 8203 (1992)
11. Y. Kakehashi: Phys. Rev. B **43**, 10820 (1991)
12. K. Fukamichi: *Amorphous Metallic Alloys*, ed. by F.E. Luborsky (Butterworths, London 1983) pp. 317–340
13. K. Fukamichi, T. Goto, H. Komatsu and H. Wakabayashi: Proc. the 4th Int. Conf. on Phys. Magn. Mater., ed. by W. Gorzkowski, H.K. Lachowicz and H. Szymczak (World Scientific, Singapore, 1989) p. 354
14. K.H.J. Buschow: Philips J. Res. Rep. **39**, 255 (1984)
15. S.N. Kaul and M. Rosenberg: Phys. Rev. B **25**, 5863 (1982)
16. A. Liénard and J.P. Rebouillat: J. Appl. Phys. **49**, 1680 (1978)
17. Y. Waseda: *The Structure of Non-Crystalline Materials: Liquids and Amorphous Solids* (McGraw-Hill, New York, 1980)
18. B.I. Min and S.J. Youn: Phys. Rev. B **49**, 9697 (1994)
19. T.T.M. Palstra, H.G.C. Werij, G.J. Nieuwenhuys, J.A. Mydosh, F.R. de Boer and K.H.J. Buschow: J. Phys. F: Met. Phys. **14**, 1961 (1984)
20. T.H. Chiang: Doctoral Thesis, Tohoku University (1993)
21. A.M. van der Kraan, K.H.J. Buschow and T.T.M. Palstra: Hyperfine Interactions **15/16**, 717 (1983)
22. L. Vegard: Z. Phys. **5**, 517 (1921)
23. G. Konczos and B. Sas: *Amorphous Metals*, ed. by H. Matyja and P.G. Zielinski (World Scientific, Singapore, 1986) pp. 105–112
24. T.H. Chiang, E. Matsubara, N. Kataoka, K. Fukamichi and Y. Waseda: J. Phys. Condens. Matter **6**, 3459 (1994)
25. C.N. Wagner, H. Ocken and M.L. Joshi: Z. Naturforsch. **20** a, 325 (1965)
26. P.A. Egelstaff, D.I. Page and J.G. Powles: Mod. Phys. **20**, 881 (1971)
27. A.H. Narten and H.A. Levy: Science **165**, 447 (1969)
28. H.A. Levy, D.M. Danford and A.H. Narten: ORNL Report (1966) No. ORNL-3960
29. D.D. Kofart, S. Nanao, T. Egami, K.M. Wong and S.J. Poon: J. Appl. Phys. **57**, 114 (1986)
30. M. Widom: *Introduction to Quasicrystals*, ed. M.V. Jarić (Academic Press, London, 1988) pp. 59–110
31. R. Yamamoto and M. Doyama: J. Phys. F **9**, 617 (1979)
32. R.B.T. Helmholdt, T.T.M. Palstra, G.J. Nieuwenhuys, J.A. Mydosh, A.M. van der Kraan and K.H.J. Buschow: Phys. Rev. B **34**, 169 (1986)
33. E. Matsubara, Y. Waseda, T.H. Chiang and K. Fukamichi: Mater. Trans. Jpn. Ins. Met. **33**, 155 (1992)
34. T.T.M. Palstra, G.J. Nieuwenhuys, J.A. Mydosh and K.H.J. Buschow: Phys. Rev. B **31**, 4622 (1985)
35. K. Fukamichi, T.H. Chiang, E. Matsubara and Y. Waseda: Sci. Rep. RITU A **41**, 9 (1995)
36. T.H. Chiang, K. Fukamichi, H. Komatsu and T. Goto: J. Phys. Condens. Matter **3**, 4055 (1991)
37. H. Wakabayashi, T. Goto, K. Fukamichi, H. Komatsu, S. Morimoto and A. Ito: J. Phys. Soc. Jpn. **58**, 3383 (1989)
38. K. Fukamichi, T.H. Chiang, N. Ohashi, N. Kataoka, E. Matsubara, Y. Waseda and T. Goto: Mater. Sci. Eng. A **181/182**, 860 (1994)

39. E.I. Gladyshevsky, O.I. Bodak and V.K. Pecharsky: *Handbook on the Physics and Chemistry of Rare-Earths*, Vol. 13, ed. by K.A.G. Schneidner Jr. and L. Eyring (North-Holland, Amsterdam, 1990) Chap. 88
40. W.A.J. Velge and K.H.J. Buschow: J. Appl. Phys. **39**, 1717 (1968)
41. A. Fujita: Doctoral Thesis, Tohoku University (1997)
42. K. Fukamichi, A. Fujita, N. Ohashi, M. Hashimoto, E. Matsubara and Y. Waseda: J. Korean Mag. Soc. **5**, 767 (1995)
43. K. Makoshi and T. Moriya: J. Phys. Soc. Jpn. **38**, 10 (1975)
44. K. Asada, A. Fujita and K. Fukamichi: (2003) to be submitted
45. K. Asada, K. Konno, M. Matsuura, M. Sakurai, A. Fujita and K. Fukamichi: J. Alloys and Compds. **350**, 47 (2003)
46. T. Moriya and A. Kawabata: J. Phys. Soc. Jpn. **35**, 669 (1974)
47. K.K. Murata and S. Doniach: Phys. Rev. Lett. **29**, 285 (1972)
48. E.C. Stoner: Phil. Mag. **15**, 1018 (1933)
49. C. Herring and C. Kittel: Phys. Rev. **81**, 869 (1951)
50. C. Herring: Phys. Rev. **85**, 1003 (1952)
51. T. Izuyama, D.J. Kim and R. Kubo: J. Phys. Soc. Jpn. **18**, 1025 (1963)
52. N.R. Bernhoeft, G.G. Lonzarich, P.W. Mitchell and D. Mck. Paul: Phys. Rev. B **28**, 422 (1983)
53. T. Moriya and A. Kawabata : J. Phys. Soc. Jpn. **34**, 639 (1973)
54. R. Kubo: J. Phys. Soc. Jpn. **12**, 570 (1957)
55. P. Rhodes and E.P. Wohlfarth: Proc. Roy. Soc. A **273**, 247 (1963)
56. J. Takeuchi and Y. Masuda: J. Phys. Soc. Jpn. **46**, 468 (1979)
57. H. Sasakura, K. Suzuki and Y. Masuda: J. Phys. Soc. Jpn. **53**, 352 (1984)
58. S.N. Kaul: J. Phys. Condens. Matter **3**, 4027 (1991)
59. S.N. Kaul and P.D. Babu: J. Phys. Condens. Matter **4**, 6429 (1992)
60. A. Fujita, T. Suzuki, K. Fukamichi, Y. Obi and H. Fujimori: Mater. Trans. JIM **36**, 852 (1995)
61. A. Fujita, K. Fukamichi, H. Aruga-Katori and T. Goto: J. Phys. Condens. Matter **7**, 401 (1995)
62. T. Moriya and Y. Takahashi: J. Phys. Soc. Jpn. **45**, 397 (1978)
63. A. Berrada, M.F. Lapierre, B. Loegel, P. Pannisod and C. Robert: J. Phys. F **8**, 845 (1987)
64. D.M. Edwards and E.P. Wohlfarth: Proc. Roy. Soc. A **303**, 127 (1968)
65. D. Wagner and E.P. Wohlfarth: Physica **112** B, 1 (1982)
66. E.P. Wohlfarth: J. Magn. Magn. Mater. **7**, 113 (1978)
67. A. Liénard, J.P. Rebouillat, P. Garoche and J.J. Veissié: J. de Phys. **41**, C8, 658 (1980)
68. H.A. Mook: J. Magn. Magn. Mater. **31–34**, 305 (1983)
69. H.G. Bohn, W. Zinn, B. Dorner and K. Kollmar: Phys. Rev. B **28**, 442 (1980)
70. H.A. Mook: Phys. Rev. Lett. **46**, 508 (1981)
71. G.G. Lonzarich and L. Taillefer: J. Phys. C **18**, 4339 (1985)
72. K. Ikeda, S.K. Dhar, M. Yoshizawa and K.A.G. Schneidner Jr.: J. Magn. Magn. Mater. **100**, 292 (1991)
73. Y. Takahashi: J. Phys. Soc. Jpn. **55**, 3533 (1986)
74. Y. Takahashi, M. Tano and T. Moriya: J. Magn. Magn. Mater. **31–34**, 329 (1983)
75. Y. Takahashi and M. Tano: J. Phys. Soc. Jpn. **51**, 1792 (1982)
76. K. Fukamichi, T. Goto and U. Mizutani: IEEE Trans. Magn. MAG-**23**, 3590 (1987)

77. T. Jarborg and M. Peter: J. Magn. Magn. Mater. **42**, 89 (1984)
78. V.L. Moruzzi, P.M. Marcus, K. Schwarz and P. Mohn: Phys. Rev. B **34**, 1784 (1986)
79. M. Podgórny and J. Goniakwaski: Phys. Rev. B **42**, 6683 (1990)
80. H. Ido, J.C. Shon, F. Pourarian, S.F. Cheng and W.E. Wallace: J. Appl. Phys. **67**, 4978 (1990)
81. K. Asada, A. Fujita, K. Fukamichi, E. Matsubara, M. Saito and Y. Waseda: J. Magn. Soc. Jpn. **23**, 480 (1999)
82. K. Fukamichi and Y. Shimada: Sci. Rep. RITU **32**, 179 (1985)
83. M. Matsui, T. Ido, K. Sato and K. Adachi: J. Phys. Soc. Jpn. **28**, 791 (1970)
84. J. Friedel: Canad. J. Phys. **34**, 1190 (1956)
85. J. Friedel: Nuovo Cimento Suppl. **2**, 287 (1958)
86. A.Z. Meńshikov, G.A. Takzei, Yu.A. Dorofeev, V.A. Kazantsev, A.K. Kostyshin and I.I. Sych: Sov. Phys. JEPT **62**, 734 (1985)
87. N.E. Brener, G. Fuster, A.J. Callaway, J.L. Fry and Y.Z. Zhao: J. Appl. Phys. **63**, 4057 (1988)
88. S. Krompiewski, U. Krey, U. Krauss and H. Ostermeier: J. Magn. Magn. Mater. **73**, 5 (1988)
89. F.J. Pinski, J. Staunton, B.L. Gyorffy, D.D. Johnson and G.M. Stocks: Phys. Rev. Lett. **56**, 2096 (1986)
90. H. Hiroyoshi and K. Fukamichi: J. Appl. Phys. **53**, 2226 (1982)
91. H. Hiroyoshi, K. Noguchi, K. Fukamichi and Y. Nakagawa: J. Phys. Soc. Jpn. **54**, 3554 (1985)
92. K. Fukamichi, H. Hiroyoshi, K. Shirakawa, T. Masumoto and T. Kaneko: IEEE Trans. Magn. MAG-**22**, 424 (1986)
93. H. Wakabayashi, K. Fukamichi, H. Komatsu, T. Goto, T. Sakakibara and K. Kuroda: Proc. Int. Symp. on Phys. of Magn. Mater., ed. by M. Takahashi, S. Maekawa, Y. Gondo and H. Nosé (World Scientific, Singapore, 1987) pp. 342–345
94. H. Wakabayashi, T. Goto, K. Fukamichi and H. Komatsu: J. Phys. Condens. Matter **2**, 417 (1990)
95. E.M. Chudnovsky, W.M. Saslow and R.A. Serota: Phys. Rev. B **33**, 251 (1986)
96. M. Fähnle and H. Kronmüller: J. Magn. Magn. Mater. **8**, 149 (1978)
97. H. Hiroyoshi, K. Fukamichi, A. Hoshi and Y. Nakagawa: Proc. Int. Conf. on High Field Magnetism, ed. by M. Date (North-Holland, Amsterdam, 1983) pp. 113–116
98. H. Kronmüller, V.A. Ignatchenko and A. Forkl: J. Magn. Magn. Mater. **134**, 68 (1994)
99. K. Fukamichi, H. Hiroyoshi, M. Kikuchi and T. Masumoto: J. Magn. Magn. Mater. **10**, 294 (1979)
100. H. Hiroyoshi, K. Fukamichi, M. Kikuchi, A. Hoshi and T. Masumoto: Phys. Lett. A **65**, 163 (1978)
101. V. Jaccarino and L.R. Walker: Phys. Rev. Lett. **15**, 258 (1965)
102. J.S. Kouvel and R.H. Wilson: J. Appl. Phys. **32**, 435 (1961)
103. M. Matsuura, H. Wakabayashi, T. Goto, H. Komatsu and K. Fukamichi: J. Phys. Condens. Matter **1**, 2077 (1989)
104. R. Harris and D. Zobin: J. Phys. F **7**, 337 (1977)
105. T.H. Chiang, K. Fukamichi and T. Goto: J. Phys. Condens. Matter **4**, 7489 (1992)

106. N. Saito, H. Hiroyoshi, K. Fukamichi and Y. Nakagawa: J. Phys. F **16**, 911 (1986)
107. M.M. Abd-Elmeguid, B. Schleede, H. Micklitz, T.T.M. Palstra, G.J. Nieuwenhuys and K.H.J. Buschow: Solid State Commun. **63**, 177 (1987)
108. Y. Kaneyoshi: *Glassy Metals*, ed. by R. Hasegawa (Boca Raton, FL, CRC, 1983) pp. 37–63
109. A. Fujita, H. Komatsu, K. Fukamichi and T. Goto: J. Phys. Condens. Matter **5**, 3003 (1993)
110. Y. Kakehashi: Phys. Rev. B **41**, 9207 (1990)
111. E.F. Wassermann: *Ferromagnetic Materials*, Vol. 5, ed. by K.H.J. Buschow and E.P. Wohlfarth (North-Holland, Amsterdam, 1990) pp. 237–322
112. T. Goto, C. Murayama, N. Mori, H. Wakabayashi, K. Fukamichi and H. Komatsu: J. de Phys. **49**, C8, 1143 (1988)
113. N. Kazama, M. Mitera and T. Masumoto: 3rd Int. Conf. Rapidly Quenched Metals (Sussex), Vol. 2, ed. by B. Cantor (The Metal Society, London, 1978) pp. 164–171
114. A. Yoshihara, Y. Shimada, T.H. Chiang and K. Fukamichi: J. Appl. Phys. **75**, 1733 (1994)
115. M.H. Grimsditch, A.P. Malozemoff and A. Brunsch: Phys. Rev. Lett. **3**, 711 (1979)
116. P. Grünberg, C.M. Mays, M. Vach and M. Grimsditch: J. Magn. Magn. Mater. **28**, 319 (1982)
117. Y. Ishikawa, K. Yamada, K. Tajima and K. Fukamichi: J. Phys. Soc. Jpn. **50**, 1958 (1981)
118. J.A. Tarvin, G. Shirane, R.J. Birgeneau and H.S. Chen: Phys. Rev. B **17**, 241 (1978)
119. J.A. Fernadez-Baca, J.W. Lynn, J.J. Rhyne and G.E. Fisher: Phys. Rev. B **36**, 8497 (1987)
120. Y. Ishikawa, S. Onodera and K. Tajima: J. Magn. Magn. Mater. **10**, 183 (1979)
121. H. Hasegawa: J. Phys. C: Solid State Phys. **14**, 2793 (1981)
122. Y. Kakehashi: J. Phys. Soc. Jpn. **50**, 1925 (1981)
123. T. Moriya and K. Usami: Solid State Commun. **34**, 95 (1980)
124. M. Shimizu: Rep. Prg. Phys. **44**, 330 (1981)
125. Y.S. Touloukian, R.K. Kirby, R.E. Taylor and P.D. Desai: *Thermal Properties of Matter*, Vol. 12, The TPRC Data Series, *Thermal Expansion: Metallic Elements and Alloys* (IFI/PLENUM, New York 1975) p. 1
126. A. Fujita and K. Fukamichi: J. Appl. Phys. **83**, 6320 (1998)
127. K. Fukamichi, T. Masumoto and M. Kikuchi: IEEE Trans. Magn. MAG-**15**, 1404 (1979)
128. H. Hiroyoshi, H. Fujimori and H. Saito: J. Phys. Soc. Jpn. **31**, 1278 (1971)
129. Y. Tanji, Y. Shirakawa and H. Moriya: J. Japan Inst. Met. **34**, 417 (1970)
130. P.E. Brommer and J.J.M. Franse: *Ferromagnetic Materials*, Vol. 5, ed. by K.H.J. Buschow and E.P. Wohlfarth (North-Holland, Amsterdam, 1990) pp. 323–396
131. G. Herzer, M. Fähnle, T. Egami and H. Kronmüller: Phys. Stat. Sol. b **101**, 713 (1980)
132. F. Acker and R. Huguenin: J. Magn. Magn. Mater. **12**, 58 (1979)
133. D. Givord, R. Lemaire, W.J. James, J.M. Moreau and J.S. Shah: IEEE Trans. Mag. MAG-**7**, 657 (1971)

134. A. Fujita, T.H. Chiang, N. Kataoka and K. Fukamichi: J. Phys. Soc. Jpn. **62**, 2579 (1993)
135. Y. Kakehashi: Phys. Rev. B **40**, 11063 (1989)
136. Y. Kakehashi: J. Phys. Soc. Jpn. **50**, 2236 (1981)
137. Y. Kakehashi: J. Magn. Magn. Mater. **103**, 78 (1992)
138. H. Hasegawa: J. Phys. Soc. Jpn. **49**, 178 (1980)
139. H. Hasegawa: J. Phys. Soc. Jpn. **49**, 963 (1980)
140. T. Moriya and H. Hasegawa: J. Phys. Soc. Jpn. **48**, 1490 (1980)
141. Y. Kakehashi: J. Phys. Soc. Jpn. **49**, 2421 (1980)
142. G. Hausch and H. Warlimont: Z. Metallk. **64**, 152 (1973)
143. K. Fukamichi, M. Kikuchi and T. Masumoto: J. Non-Cryst. Solids **61/62**, 961 (1984)
144. A. Kussmann, M. Auwarter and G.G. von Rittlerg: Ann. Phys. **4**, 174 (1948)
145. M. Shiga, K. Makita, K. Uematsu, Y. Muraoka and Y. Nakamura: J. Phys. Condens. Matter **2**, 1239 (1990)
146. S. Ito, K. Aso, Y. Makino and S. Uedaira: Appl. Phys. Lett. **37**, 665 (1980)
147. H. Tange, Y. Tanaka, M. Goto and K. Fukamichi: J. Magn. Magn. Mater. **81**, L243 (1989)
148. K. Fukamichi, T.H. Chiang, A. Fujita, H. Tange, S. Kawabuchi and T. Ono: J. Phys. Condens. Matter **7**, 2875 (1995)
149. E.R. Callen and H.B. Callen: Phys. Rev. **120**, 578 (1963)
150. E.R. Callen and H.B. Callen: Phys. Rev. A **139**, 455 (1965)
151. R.C. O'Handley: Phys. Rev. B **18**, 930 (1978)
152. B.S. Berry and W.C. Pritchet: Solid State Commun. **26**, 827 (1978)
153. R.C. O'Handley: Solid State Commun. **21**, 1119 (1977)
154. M. Shiga, *Electronic and Magnetic Properties of Metals and Ceramics*, Vol. 3B, Part II, ed. by K.H.J. Buschow (VCH Pub. Inc. Weinheim, 1993) pp. 159–210
155. H. Tange, T. Kamimori, M. Goto and K. Fukamichi: J. Magn. Magn. Mater. **90/91**, 335 (1990)
156. M.H. Grimsditch: *Light Scattering in Solids V*, ed. by M. Cardona and G. Güntherodt (Springer, Berlin, 1988) Chap. 7 and references therein
157. P. Grünberg: Prog. Surf. Sci. **18**, 1 (1985)
158. A. Tomizuka, H. Iwasaki, K. Fukamichi and T. Kikegawa: J. Phys. F **14**, 1507 (1984)
159. T.H. Chiang, H. Komatsu, A. Matsunaga, K. Fukamichi, A. Yoshihara, Y. Shimada and T. Goto: Mater. Sci. Eng. A **181/182**, 958 (1994)
160. K. Shirakawa, K. Fukamichi, T. Kaneko and T. Masumoto: Physica B&C **119**, 192 (1983)
161. K. Fukamichi, K. Shirakawa, T. Kaneko and T. Masumoto: Proc. 5th Int. Conf. on Rapidly Quenched Metals, ed. by S. Steeb and H. Warlimont (North-Holland, Amsterdam, 1985) pp. 1165–1168
162. K. Fukamichi, Y. Satoh, K. Shirakawa, T. Masumoto and T. Kaneko: J. Magn. Magn. Mater. **54–57**, 231 (1986)
163. K. Fukamichi, K. Shirakawa, Y. Satoh, H. Komatsu, T. Masumoto and T. Kaneko: High Pressure Research **1**, 193 (1989)
164. H. Tange and M. Goto: J. Phys. Soc. Jpn. **49**, 957 (1980)
165. M. Shiga, Y. Muraoka and Y. Nakamura: J. Magn. Magn. Mater. **10**, 280 (1979)

166. J.L. Snoek: Physica **4**, 8532 (1937)
167. H. Tange and T. Tokunaga: J. Phys. Soc. Jpn. **27**, 789 (1969)
168. J.R.L. de Almeida and D.J. Thouless: J. Phys. A **11**, 983 (1978)
169. M. Gabey and G. Toulouse: Phys. Rev. Lett. **47**, 201 (1981)
170. Y. Kakehashi: Physica B **161**, 143 (1989)
171. K. Fukamichi, T.H. Chiang and T. Goto: J. Phys. Condens. Matter **9**, 6069 (1997)
172. J. Inoue and M. Shimizu: Phys. Lett. A **90**, 85 (1982)
173. N. Buis, J.J.M. Franse and G. Hilscher: Physica B **86–88**, 319 (1977)
174. Y. Nakamura, M. Hayase, M. Shiga, Y. Miyamoto and N. Kawai: J. Phys. Soc. Jpn. **30**, 720 (1971)
175. W. Ludorf, M.M. Abd-Elmeguid and H. Micklitz: J. Man. Magn. Mater. **78**, 171 (1989)
176. I.V. Medvedeva, A.A. Ganin, Ye.V. Sheherbakova, A.S. Yermolenko and Yu.S. Bersenev: J. Alloys and Compd. **178**, 403 (1992)
177. T. Moriya: J. Magn. Magn. Mater. **14**, 1 (1979)
178. N.D. Lang and H. Ehrenreich: Phys. Rev. **168**, 605 (1968)
179. H. Tange, T. Yonei and M. Goto: J. Phys. Soc. Jpn. **50**, 454 (1981)
180. J.J.M. Franse: J. Magn. Magn. Mater. **10**, 259 (1979)
181. H. Tange, Y. Tanaka, T. Kamimori and M. Goto: J. de Phys. (Paris) **49**, C8, 1283 (1988)
182. H. Tange, Y. Tanaka and K. Shirakawa: J. de Phys. (Paris) **49**, C8, 1281 (1988)
183. M. Shimizu: J. Magn. Magn. Mater. **20**, 47 (1980)
184. E.P. Wohlfarth: Physica B **91**, 305 (1977)
185. E.P. Wohlfarth: Solid State Commun. **35**, 797 (1980)
186. International Tables for X-Ray Crystallography, Vol. III (Kynoch, Birmingham UK 1968) p. 250
187. International Tables for X-Ray Crystallography, Vol. IV (Kynoch, Birmingham UK 1974) p. 99

Index

Printed in the United States
99605LV00003B/21/A